An Atlas of
INDUCED SPUTUM
An Aid for Research and Diagnosis

THE ENCYCLOPEDIA OF VISUAL MEDICINE SERIES

An Atlas of
INDUCED SPUTUM
An Aid for Research and Diagnosis

Ratko Djukanovic, MD, DM, FRCP
Division of Infection, Inflammation and Repair,
Respiratory Cell and Molecular Biology
Southampton General Hospital
Southampton
United Kingdom

and

Peter J. Sterk, MD, PhD
Department of Pulmonology
Leiden University Medical Center
Leiden
The Netherlands

The Parthenon Publishing Group
International Publishers in Medicine, Science & Technology

A CRC PRESS COMPANY
BOCA RATON   LONDON   NEW YORK   WASHINGTON, D.C.

Published in the USA by
The Parthenon Publishing Group
345 Park Avenue South
New York, NY 10010, USA

Published in the UK by
The Parthenon Publishing Group
23–25 Blades Court
Deodar Road
London, SW15 2NU, UK

Library of Congress Cataloging-in-Publication Data
An atlas of induced sputum : an aid for research and diagnosis / [edited by] Ratko
    Djukanovic and Peter J. Sterk.
    p. ; cm. -- (The encyclopedia of visual medicine series)
    Includes bibliographical references and index.
    ISBN 1-84214-005-1 (alk. paper)
    1. Sputum--Examination--Atlases. 2. Lungs--Diseases--Cytodiagnosis--Atlases. 3.
    Exfoliative cytology--Atlases. I. Title: Induced sputum. II. Sterk, Peter J. III. Djukanovic,
    Ratko. IV. Series.
    [DNLM: 1. Sputum--cytology--Atlases. 2. Lung Diseases--diagnosis--Atlases. QY 17
    A8813 2003]
    RB51.A87 2003
    616.2'407582--dc22
                                                                      2003061738

British Library Cataloguing in Publication Data
Djukanovic, Ratko
    An atlas of induced sputum : an aid for research and
    diagnosis. – (The encyclopedia of visual medicine series)
    1. Sputum – Examination
    I. Title II. Sterk, Peter
    612.3'13

ISBN 1-84214-005-1

Typeset by Siva Math Setters, Chennai, India
Printed and bound by T. G. Hostench S. A., Spain

Contents

List of principal contributors

Elena Bacci
Dipartimento Cardio-Toracico
Ospedale di Cisanello
via Paradisa 2
56100 Pisa
Italy

Pascal Chanez
Clinique des Maladies Respiratoires
Hôpital Arnaud de Villeneuve
371 Av du Doyen Gaston Giraud
F-34295-Montpellier Cedex 5
France

Ulrich Costabel
Department of Pneumology and Allergy
Ruhrlandklinik
Tüschener Weg 40
D-45239 Essen
Germany

Ratko Djukanovic
Division of Infection, Inflammation and Repair
Respiratory Cell and Molecular Biology
Mailpoint 810
Level D, Centre Block
Southampton General Hospital
Southampton SO16 6YD
UK

Ann Efthimiadis
Firestone Institute for Respiratory Health
St. Joseph's Healthcare and McMaster University
50 Charlton Ave. E.
Hamilton ON L8N 4A6
Canada

Elizabeth Fireman
Laboratory of Pulmonary and Allergic Diseases
National Laboratory Service for Interstitial Lung
  Diseases

Tel-Aviv Medical Center
6 Weizman Street
Tel-Aviv 64239
Israel

Peter G. Gibson
School of Medical Practice
Faculty of Health
University of Newcastle
  and
Department of Respiratory and Sleep Medicine
Hunter Medical Research Institute
Level 3, John Hunter Hospital
Hunter Region Mail Centre
NSW 2310
Australia

Ruth H. Green
Department of Respiratory Medicine and Thoracic
  Surgery
University Hospitals of Leicester NHS Trust
Glenfield Hospital
Groby Road
Leicester
LE3 9QP
UK

Qutayba Hamid
Professor of Medicine and Pathology
McGill University
Meakins Christie Labs
Montreal, Quebec H2X 2P2
Canada

Margaret Kelly
Department of Pathology and Molecular Medicine
Hamilton Health Sciences
Room 4H8
McMaster University
1200 Main St West

Hamilton ON
L8N 3Z5
Canada

Renaud A. Louis
Department of Pneumology
CHU Sart-Tilman
University of Liege
Liege 4000
Belgium

Piero Maestrelli
Department of Environmental Medicine and Public
  Health
Laboratory of Lung Pathophysiology
University of Padova
via Giustiniani 2
35128 Padova
Italy

Cristina E. Mapp
Department of Clinical and Experimental Medicine
Section of Hygiene and Occupational Medicine
University of Ferrara
Via Fossato di Mortara 64/b
44100 Ferrara
Italy

Pierluigi Paggiaro
Dipartimento Cardio-Toracico
Ospedale di Cisanello
via Paradisa 2
56100 Pisa
Italy

Debbie Parker
Department of Respiratory Medicine and Thoracic
  Surgery
University Hospitals of Leicester NHS Trust
Glenfield Hospital
Groby Road
Leicester
LE3 9QP
UK

Ian D. Pavord
Department of Respiratory Medicine and Thoracic
  Surgery
University Hospitals of Leicester NHS Trust
Glenfield Hospital
Groby Road
Leicester
LE3 9QP
UK

Paula Rytilä
Department of Allergy
Helsinki University Central Hospital
Post Box 160
00029 HUCH
Helsinki
Finland

Janis Shute
School of Pharmacy and Biomedical Sciences
University of Portsmouth
White Swan Road
Portsmouth PO1 2DT
UK

Jodie L. Simpson
Department of Respiratory and Sleep Medicine
Hunter Medical Research Institute
Level 3
John Hunter Hospital
Locked Bag 1
Hunter Mail Exchange
NSW 2310
Australia

Antonio Spanevello
Fondazione Salvatore Maugeri
Institute of Care and Research
Via per Mercadante KM 2
70020 Cassano Murge (Bari)
Italy

Peter J. Sterk
Lung Function Lab, C2–P
Leiden University Medical Center
Albinusdreef 2
P.O. Box 9600
Leiden NL–2300 RC
The Netherlands

Peter Wark
Brooke Laboratories
RCMB Research Division
Southampton General Hospital
Tremona Rd
Southampton SO16 6YD
UK

Introduction

Ratko Djukanovic and Peter J. Sterk

The ability to understand disease mechanisms and make a correct diagnosis is greatly enhanced by analyzing tissue or body fluid samples; this information complements that obtained by functional tests. This principle has long been applied in nephrology, hepatology and dermatology but was not fully accepted until recently in respiratory medicine. Thus, asthma and chronic obstructive pulmonary disease (COPD) were diagnosed on the basis of history and lung function tests, and the study of pathogenesis was limited by the relative lack of insight into the pathological changes in the airways and parenchyma. In the mid 1980s, first attempts to study airway pathology were made using bronchoscopy. This tool has provided extremely valuable insight into the pathology of asthma, and more recently COPD, but the invasive nature of the technique has meant that its use was limited mainly to academic centres with appropriate facilities and significant expertise to undertake a procedure that is not without risk.

It has long been appreciated that sputum provides insight into mechanisms of diseases of the lungs. Hippocrates identified sputum as one of the four essential humors in the body. Ever since the development of cytology and microbiology, sputum has been used to diagnose cancer and respiratory infections, including tuberculosis. In the last century Gollasch observed an increase in eosinophils in the sputum of asthmatics. In 1958 Dr Morrow Brown recognized the value of assessing sputum for the presence of eosinophils, which predicted which patients would respond to corticosteroids. After a period during which the use of induced sputum to assess airway diseases was almost forgotten, refinements in methodology have progressed significantly in the last 10–15 years to the point that today it is one of the more referenced methods used to assess airway inflammation in research and clinical practice. Recently, a task force of the European Respiratory Society (ERS) investigated the value of sputum induction and guidelines were provided for how sputum should be induced, processed and interpreted. These can be found in a supplement of the European Respiratory Journal, 2002, volume 37.

The aim of this Atlas is to be a useful reference book, with many illustrations and easy-to-read text. We wish to help researchers, clinicians, laboratory technicians, and all others involved in clinical trials and drug development to acquire insight into some of the major advances in the use of sputum to study lung disease, excluding lung cancer since this is usually and better covered by lung pathologists. The Atlas has been put together by experts in the field of induced sputum, who have selected the most relevant and exciting data and have presented them in a way that is easy to read and allows easy recognition of the potential of this method.

Sputum is defined as secretions that are expectorated from the lower respiratory tract. As such, it is composed of a fluid phase, which contains abundant mucins, and a cellular phase, which contains inflammatory cells and epithelial cells that are shed into the airway lumen. The Atlas covers the important topics of methods and safety of induction, the principles of processing and analysis of the fluid and cellular phases of sputum, and the exciting application of the technique to study a variety of lung diseases.

Ratko Djukanovic
Peter J. Sterk

CHAPTER 1

Sputum induction: methods and safety

Pierluigi Paggiaro, Antonio Spanevello and Elena Bacci

The aim of sputum induction is to collect an adequate sample of lower airway secretions in patients who are not able to produce sputum spontaneously. Sputum induction requires a high degree of patient cooperation and is best conducted in a quiet, secluded area to minimize embarrassment for the patient (Figure 1.1). Adequate facilities and equipment are needed (Table 1.1, Figure 1.2). Infection control procedures for the protection of personnel and patients must be carried out according to the local anti-infection policy, which sometimes includes purpose-built chambers (Figure 1.1).

For successful induction it is essential that a sufficiently high output of saline aerosol with an adequate particle size is achieved. This is best done with ultrasonic nebulizers, the output of which needs to be accurately determined and set at the same level for a given study. Fresh sterile saline solution should be used. There is no evidence to suggest that larger volumes of saline are more successful than smaller ones, and the consensus at present is that an output of approximately 1 ml/min is sufficient to achieve a high success rate.

Inhalation of isotonic or hypertonic solutions administered by nebulization induces small amounts of airway secretions that can be expectorated more or less easily (Figure 1.3). Some patients, such as those with acute asthma exacerbations or severe current symptoms, and patients with chronic obstructive pulmonary disease

(COPD) or cystic fibrosis, can produce significant amounts of sputum spontaneously, i.e. without the need for inhalation of saline. This can be sufficient for some studies in that spontaneous sputum contains similar percentages of inflammatory cells and amounts of some mediators as induced sputum (Figure 1.4), albeit at the expense of cell viability and the quality of samples. However, the majority of patients need to inhale saline before they can produce an adequate sample.

The technique of sputum induction has been standardized to enable comparison between researchers and to ensure adequate reproducibility of the technique. Frequent repetition of sputum induction can itself lead to airway inflammation, resulting in a change in cell populations obtained by subsequent inductions. Thus, repeating induction 8–24 h after an initial induction can cause an increase in neutrophils (Figure 1.5), and it is possible that mediator levels also change. For this reason, an interval of 2 days between inductions is recommended.

SAFETY

As with all procedures, safety of sputum induction is paramount. The risk of excessive bronchoconstriction caused by sputum induction in subjects with susceptible airways cannot be underestimated. Predictors of excessive bronchoconstriction

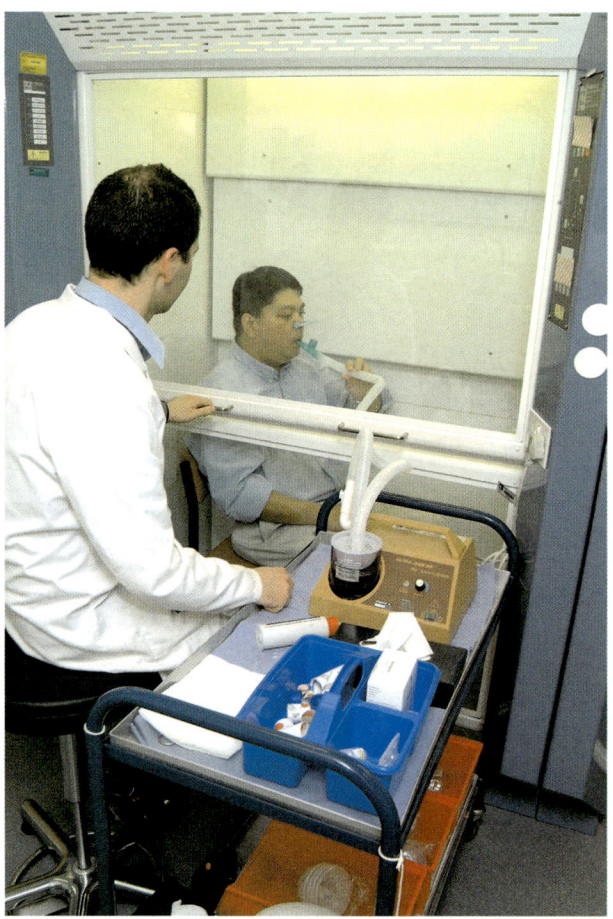

Figure 1.1 Sputum induction carried out in a volunteer participating in a research project. The subject is wearing a nose-clip to avoid breathing through the nose and thus reduce contamination of sputum by nasal secretions. A technician or nurse must be present throughout the procedure, measuring peak exploratory flow using a hand-held device (Wright's mini peak flow meter shown)

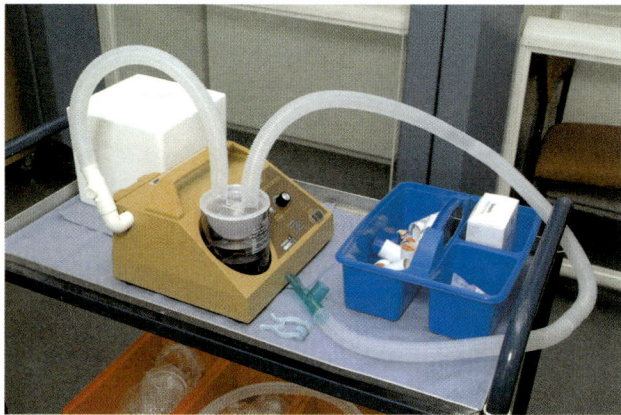

Figure 1.2 Trolley with an ultrasonic nebulizer. The sputum sample is expectorated into a Petri dish and placed on ice in the white polystyrene box

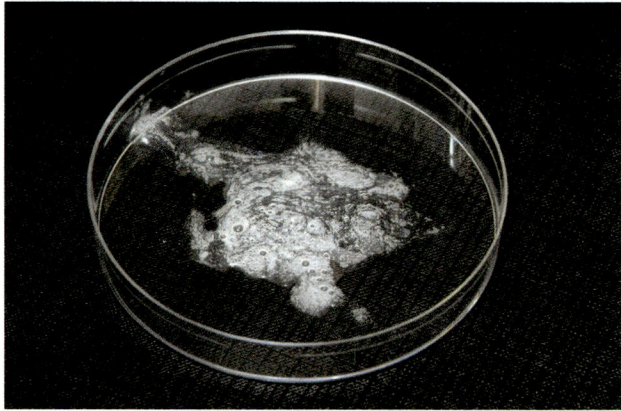

Figure 1.3 Sputum sample expectorated and collected in a Petri dish over a period of 15 min. A good sample contains small amounts of saliva only, with the majority of the sample being viscous due to the presence of high-molecular-weight mucins that form the gel phase of sputum

Table 1.1 List of facilities and instruments required for sputum induction

Facilities:
- quiet environment
- sterile saline solutions

Instruments:
- ultrasonic nebulizer
- spirometer
- safety equipment (resuscitation equipment, oxygen supply, rescue medications)

reported so far are the degree of baseline airflow limitation and the degree of airway hyperresponsiveness to methacholine or histamine. However, this has not been universally confirmed and measurement of hyperresponsiveness does not remove the need for vigilance. Studies have reported a strong correlation between recent overuse of a short-acting β_2-agonist and the magnitude of fall in forced expiratory volume (FEV_1) after sputum induction, and continuous β_2-agonist

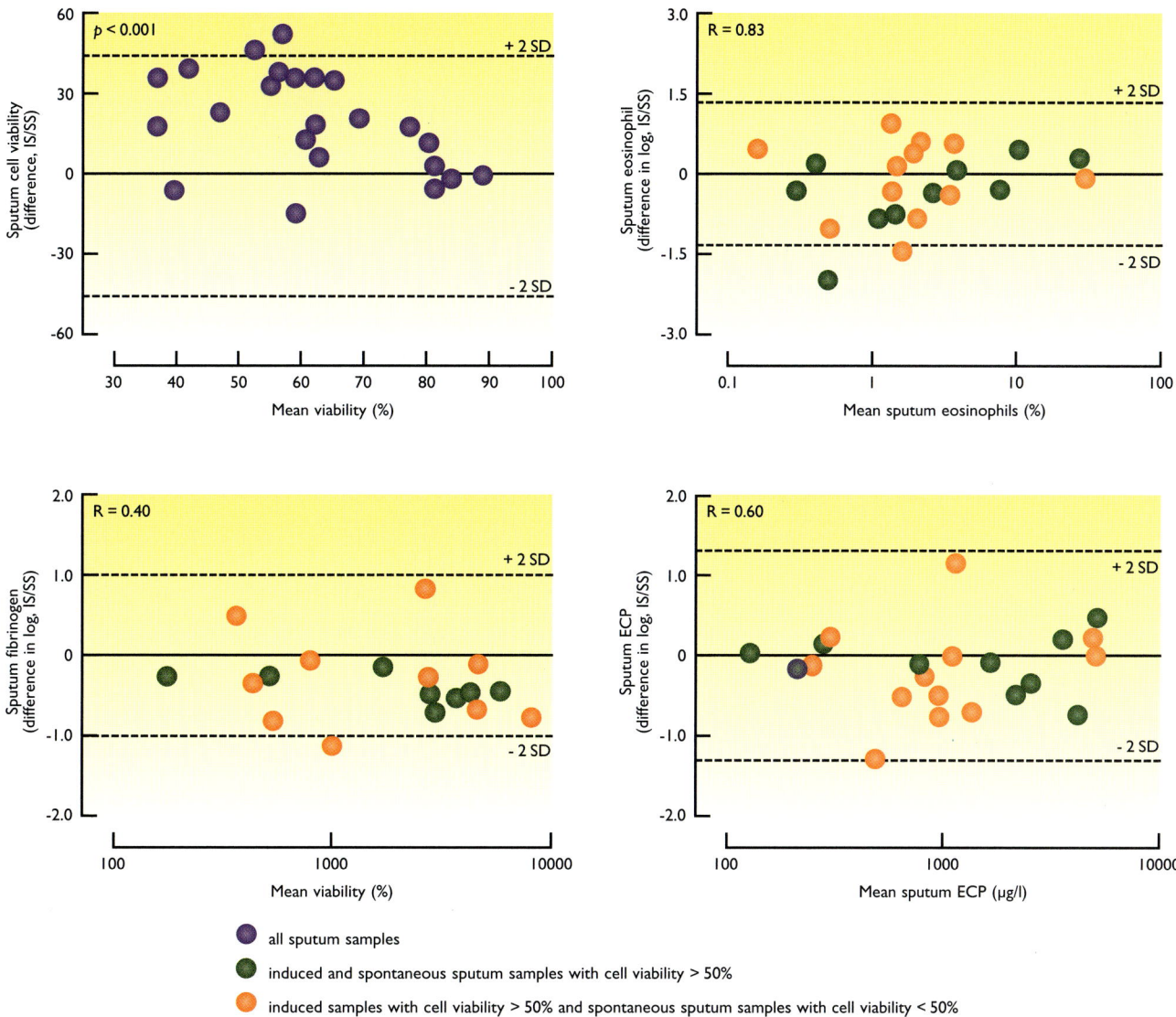

Figure 1.4 Comparison of sputum obtained spontaneously with sputum obtained by induction using hypertonic saline in respect of cell viability, eosinophil counts, concentration of fibrinogen, and eosinophil cationic protein (ECP). The difference (in logs, except for cell viability) between induced (IS) and spontaneous (SS) sputum is plotted against the mean of the two values. R is the intraclass correlation coefficient and the marked area is ± 2 standard deviations (SD) of the mean of the two differences. Adapted with permission from Pizzichini MM, Popov TA, Efthimiadis A, *et al*. Spontaneous and induced sputum to measure indices of airway inflammation in asthma. *Am J Respir Crit Care Med* 1996;154:866–9

use can result in a reduction of the bronchoprotective effect against a variety of both specific and non-specific bronchoconstrictor stimuli. It has also been demonstrated that 200 or 400 µg of salbutamol does not protect against excessive bronchoconstriction if there is exposure to a relatively strong bronchoconstrictive stimulus.

In view of these concerns, sputum induction should be conducted by an experienced technician or nurse and supervised by an experienced physician. This is needed for early and dependable recognition of bronchoconstriction in patients with hyperreactive airways. Full resuscitation equipment and additional rescue bronchodilator

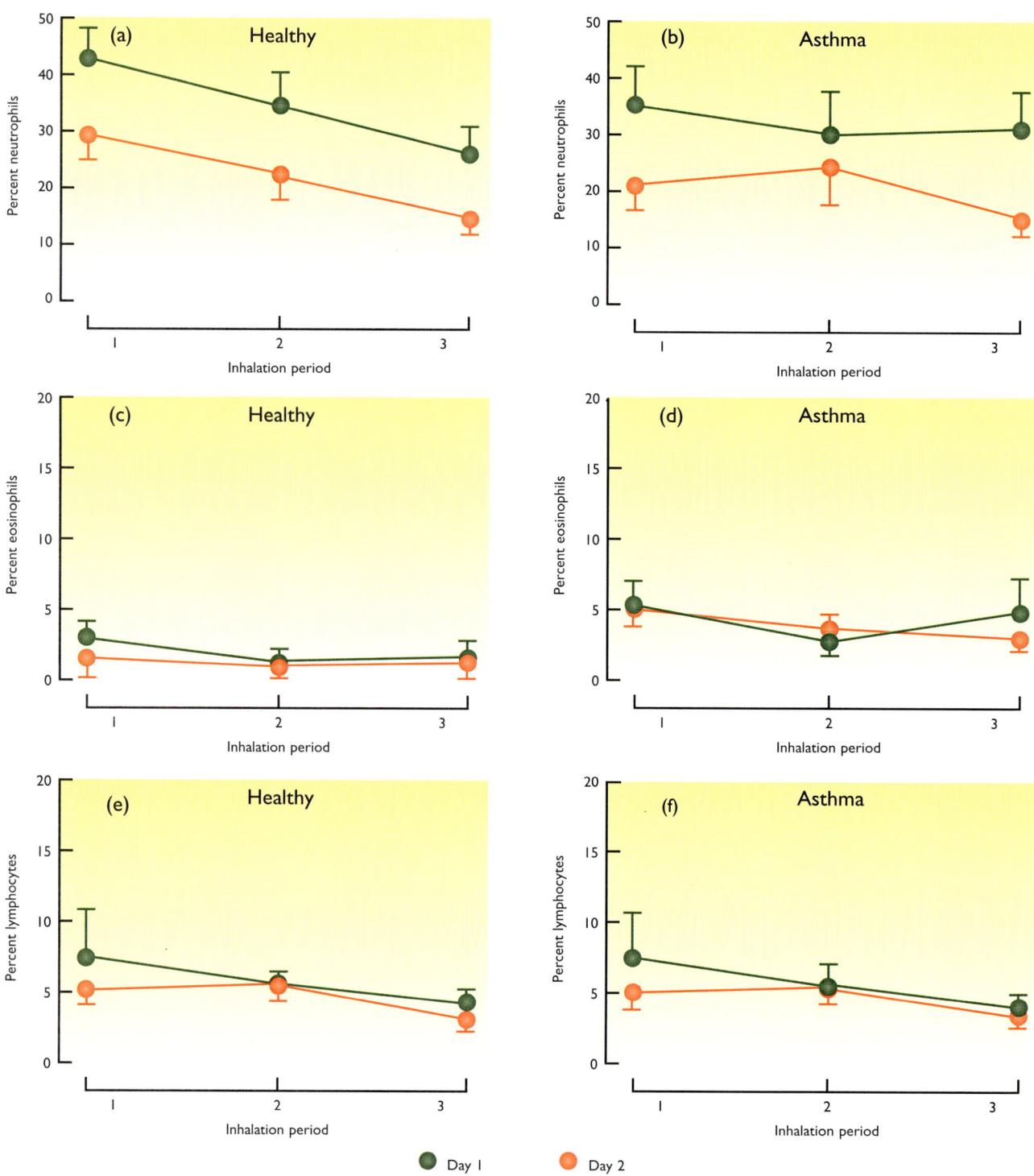

Figure 1.5 The effect of repeat sputum induction on sputum cell counts. Mean (SE) percentage of neutrophils (a, b), eosinophils (c, d), and lymphocytes (e, f) measured on day 1 and 24 h later on day 2 are shown. Inhalation periods 1–3 indicate consecutive 10-min sampling periods within each of the two sputum inductions. The increase in levels of neutrophils in the second sputum induction, compared with the first, was statistically significant in both healthy and asthmatic subjects ($p < 0.01$). Adapted with permission from the BMJ Publishing Group from Holz O, Richter K, Jorres RA, *et al.* Changes in sputum composition between two inductions performed on consecutive days. *Thorax* 1998;53:83–6

medication together with any other emergency drugs must be immediately available. Oxygen saturation should be monitored if there is any concern about resting hypoxemia. Supplemental oxygen must be immediately available.

Based on considerable experience accrued over the past 10 years, recommendations have been made for how sputum induction should be conducted. The method varies depending on whether the investigator sees the patient as being at low risk of bronchoconstriction – in which case a standard protocol is applied – or at high risk, when an alternative protocol is used (Table 1.1).

Pretreatment with a short-acting β_2-agonist, salbutamol (200 µg), delivered via a standard metered dose inhaler is generally used in all cases. Higher doses are not recommended, even in severe asthmatics, since this may reduce the effectiveness of any additional doses that may be needed during the course of induction. Pretreatment has been shown to have no effect on the numbers of various inflammatory cell types (Figure 1.6) and some mediators such as eosinophil cationic protein (ECP), but it reduces histamine levels. The effect on other soluble mediators is unknown.

Bronchodilator pretreatment should always be given except in rare (research) cases when the inhalation of hypertonic saline is part of a bronchial challenge protocol to assess airway responsiveness. However, investigators should be aware of the fact that a greater degree of bronchoconstriction is likely to occur in the absence of pretreatment.

MONITORING LUNG FUNCTION – AN IMPORTANT SAFETY MEASURE

So far there has been a large variation in protocols for pulmonary function monitoring during sputum induction. Most investigators have used spirometry, although some have used peak flow meters. Spirometers are preferred because of the higher sensitivity of FEV_1. Spirometry should be carried out before the procedure to assess the baseline airway caliber and ensure that it is safe to proceed. Repeat measurements during induction are needed to avoid excessive bronchoconstriction.

Measurement of FEV_1 is conducted before and after 10-min administration of the pre-induction bronchodilator. The inhalation of saline can then begin. Since dyspnea is poorly perceived by some individuals and bronchospasm may occur early during inhalation, it is best to measure pulmonary function within the first minute of nebulization in order to detect subjects who are very sensitive to hypertonic saline. Thereafter, it is measured after each inhalation period (Table 1.1) or if the procedure is interrupted due to breathlessness or wheeze. If FEV_1 falls to 20% of baseline the procedure should be aborted and further β_2-agonist given. In such cases, it is advised that the FEV_1 should return to within 5% of baseline before recommencing saline inhalation. At the end of the induction procedure, patients should be given additional short-acting β_2-agonist, particularly if there has been a fall in FEV_1 of more than 10% from the baseline value. Patients should remain and be monitored in the laboratory until their FEV_1 has returned to within 5% of the baseline value.

The recommendations of the European Respiratory Task Force on Induced Sputum are summarized in Table 1.2.

CONCENTRATION OF SALINE SOLUTION AND DURATION OF INHALATION

In studies conducted so far, there has been significant variation in the concentrations of saline used for sputum induction, ranging from 0.9 to 7% saline. There is now a consensus for induction to be conducted with 4.5% saline solution as standard because it is commercially available, effective and generally well tolerated. However, as shown in Table 1.2, other concentrations can also be used.

In an attempt to reduce the risk of excessive bronchoconstriction caused by hypertonic saline, some investigators gradually change concentrations during the procedure, starting with 3% and increasing to 4 and 5%. Hypertonic saline is

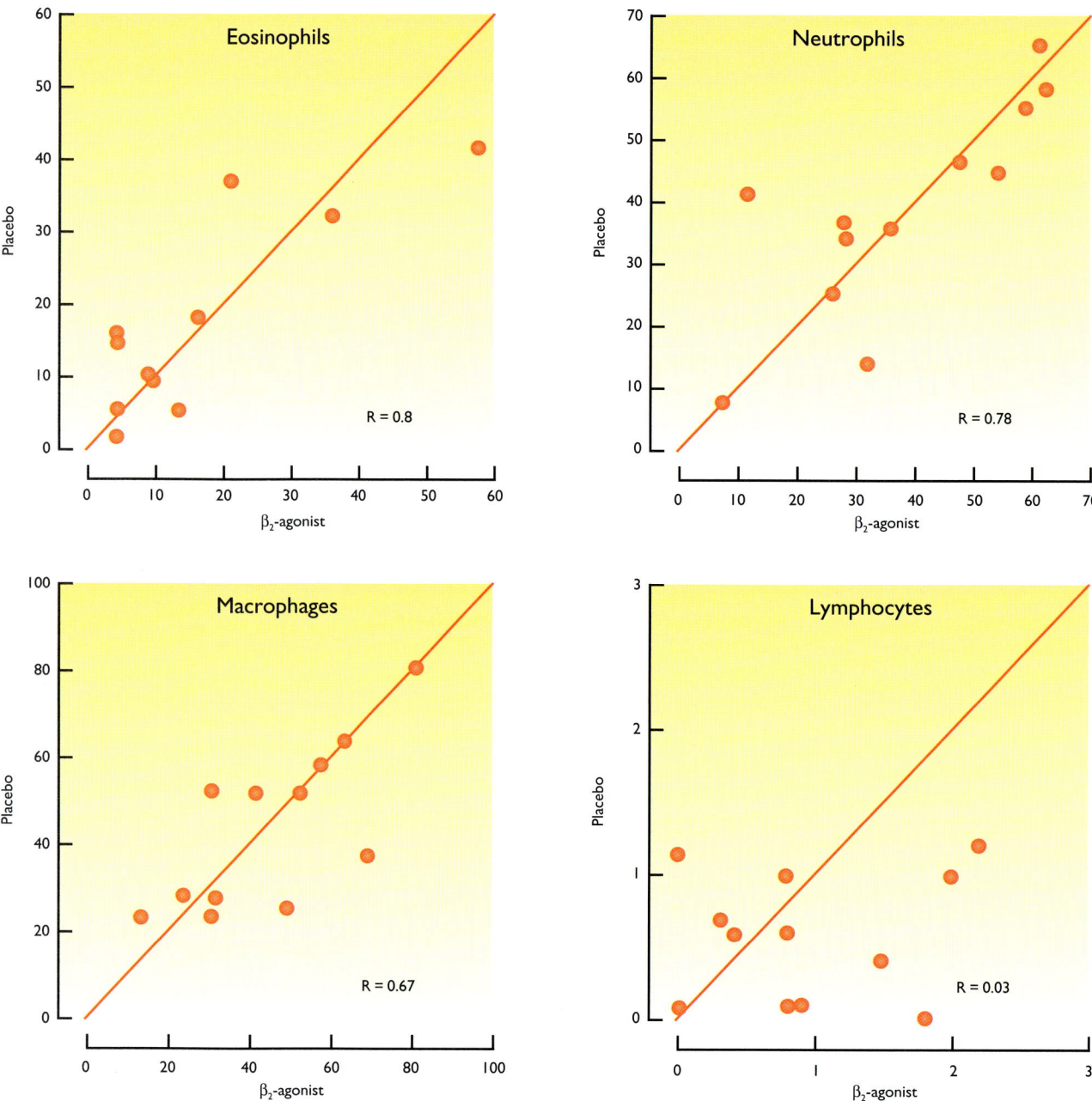

Figure 1.6 Effect of pretreatment with salbutamol compared with placebo on differential cell counts in induced sputum obtained on two separate days. The figure shows identity plots and R values (intraclass correlation coefficients). The cell counts for eosinophils, neutrophils and macrophages are similar for the two days, suggesting little effect of pretreatment with salbutamol; the difference seen for lymphocytes probably reflects generally poor reproducibility of lymphocyte counts in sputum. Adapted with permission from Cianchetti S, Bacci E, Ruocco L, et al. Salbutamol pretreatment does not change eosinophil percentage and eosinophilic cationic protein concentration in hypertonic saline-induced sputum in asthmatic subjects. *Clin Exp Allergy* 1999;29:712–18

Table 1.2 Methods for sputum induction. Reproduced with permission from Pizzichini E, Pizzichini MM, Leigh R, Djukanovic R, Sterk PJ. Safety of sputum induction. *Eur Respir J* (Suppl) 2002;37:9s–18s

Standard method (in high-risk patients use alternative procedure)

- Give detailed information and clear instructions to the patient prior to the procedure
- Check safety equipment and set up ultrasonic nebulizer (output approx 1 ml/min)
- Measure pre-bronchodilator forced exiratory volume (FEV_1)
- Administer 200 μg of inhaled salbutamol before commencing
- After 10 min measure post-bronchodilator FEV_1
- Use either a fixed concentration of sterile saline solution (e.g. 3 or 4.5%) or incremental concentrations of saline (3, 4 and 5%)
- Perform induction in 5-min intervals for no longer than 20 min. Alternatively, induction can be conducted at 1, 4, and 5 min, with three further 5-min periods
- Measure FEV_1 at the end of each induction interval*
- Ask the patient to cough and spit at the fifth, tenth, fifteenth and twentieth minute of induction or whenever they get the urge to do so

Alternative method (in high-risk patients)

- Give detailed information and clear instructions to the patient prior to the procedure
- Check safety equipment and set up ultrasonic nebulizer (output approx 1 ml/min)
- Measure pre-bronchodilator FEV_1
- Administer 200 μg of inhaled salbutamol
- After 10 min measure post-bronchodilator FEV_1
- Start with 0.9% NaCl sterile saline solution and perform induction for 30 sec, 1 min, and 5 min, measuring FEV_1 after each period as a safety precaution. If this fails to induce sputum, increase the saline concentration to 3%, induce for 30 sec, 1 min, and 2 min. If this also fails to induce sputum, increase further to 4.5% and induce for 30 sec, 1 min, 2 min, 4 min and 8 min
- If normal saline is successful in inducing sputum, there is no need to progress to higher concentrations. The same applies for 3% saline
- Measure FEV_1 at the end of each induction interval*
- Induction must be stopped if FEV_1 falls by 20% of baseline
- If the patient does not cough spontaneously, ask them to attempt to cough and spit after the 4- and 8-min periods

*Stop induction if there is a fall in FEV_1 of more than 20% compared with the post-bronchodilator value or if symptoms occur

slightly more effective than isotonic saline in inducing sputum, and importantly there is no difference in the cell composition of sputum induced with isotonic or hypertonic saline (Figure 1.7). However, the effect of varying concentration on levels of soluble mediators in induced sputum is not known.

The duration of inhalation is a very important variable in sputum induction since both cellular and biochemical components of induced sputum change during the course of induction (Figure 1.8) (also see Chapter 3). Neutrophils and eosinophils are prominent in samples collected early during sputum induction, whereas lymphocyte and macrophage counts increase in latter samples, suggesting that different compartments of the lung are sampled at different time-points during induction, i.e. central airways are sampled early, whereas peripheral airways and alveoli are sampled later. The maximum acceptable duration of induction has not been formally studied, but it is very important to keep the duration of inhalation constant between subjects, and especially in the same subject, to enable comparison. Shorter inhalation times (15–20 min) appear to have similar success rates and feasibility as longer inhalation times (30 min); therefore, the consensus is that the duration should be between 15 and 20 min.

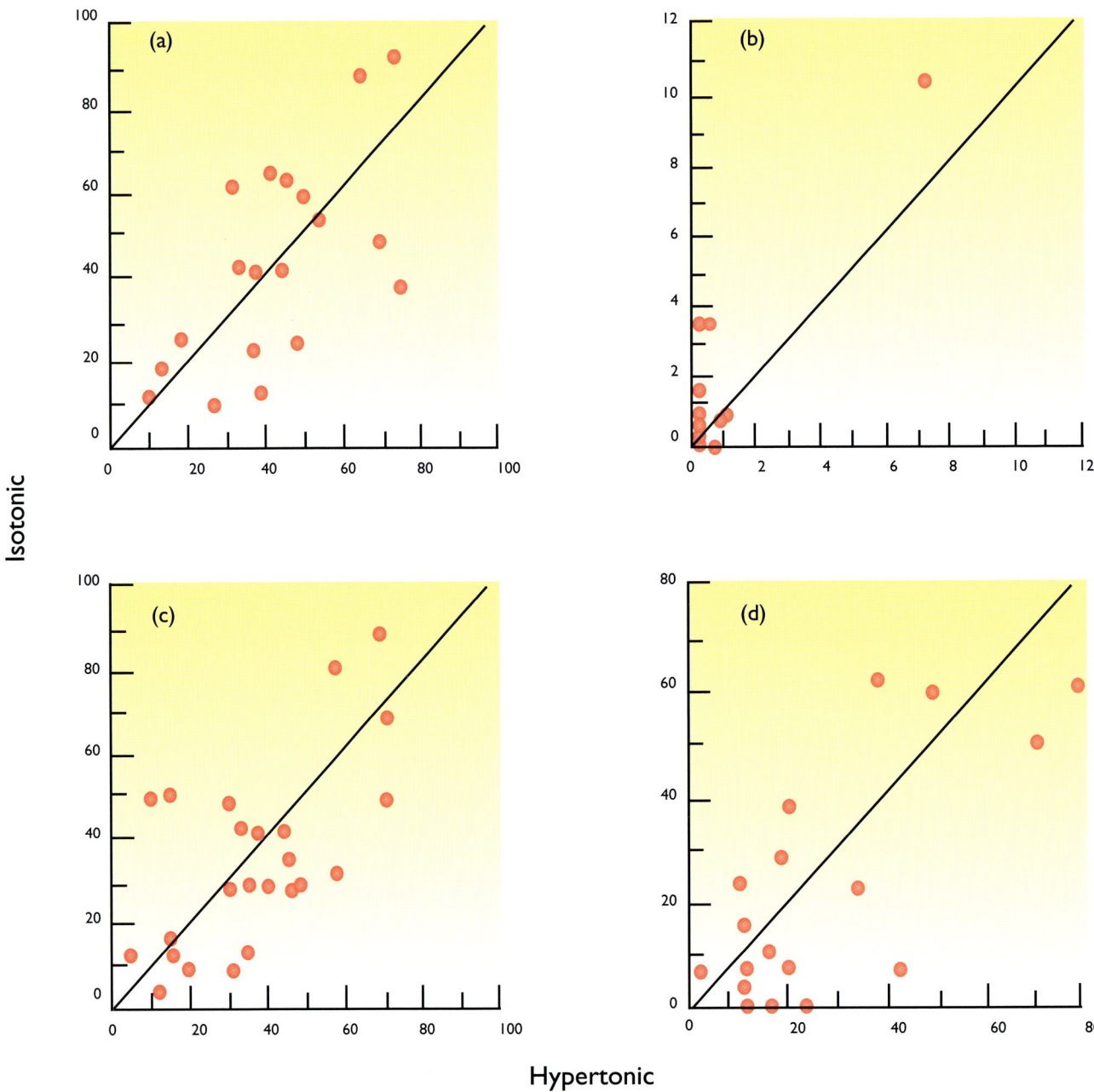

Figure 1.7 Comparison of differential cell counts in sputum samples obtained in the same individuals using isotonic (normal) and hypertonic saline for induction. The intraclass correlation coefficients are: (a) +0.505 for macrophages; (b) +0.544 for lymphocytes; (c) +0.644 for neutrophils; and (d) +0.642 for eosinophils. This suggests little difference in cell counts between spontaneous and induced sputum. Adapted with permission from Bacci E, Cianchetti S, Paggiaro PL, *et al.* Comparison between hypertonic and isotonic saline-induced sputum in the evaluation of airway inflammation in subjects with moderate asthma. *Clin Exp Allergy* 1996;26:1395–400

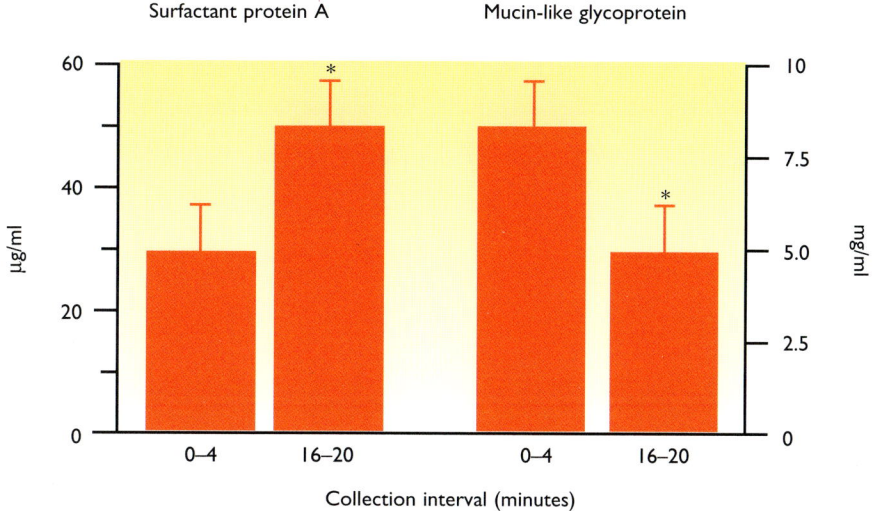

Figure 1.8 Changes in the levels of surfactant protein A and mucin-like glycoprotein in induced sputum samples during the course of sputum induction. Samples were collected individually at the beginning (0–4 min) and end (16–20 min) of a 20-min sputum induction. *Significantly different from the 0–4-min interval ($p < 0.05$). Reproduced with permission from Gershman NH, Liu H, Wong HH, *et al.* Fractional analysis of sequential induced sputum samples during sputum induction: evidence that different lung compartments are sampled at different time points. *J Allergy Clin Immunol* 1999;104:322–8

EXPECTORATION TECHNIQUE

Patients may be asked to stop inhalation at set intervals in order to cough up sputum (e.g. every 5 min) or to stop only when they feel the urge to cough. There is nothing to suggest that this changes the result. Spitting saliva into one cup before coughing sputum into another has been shown to decrease the percentage of squamous cells in sputum (whole expectorate) by 30% and increase the concentration of ECP in the supernatant by 80%. Other authors have recommended, but not shown clear benefits of, the use of various steps such as washing the mouth with water, blowing the nose prior to the procedure, or wearing a nose-clip.

CHAPTER 2

Analysis of sputum cells: cytology, immunocytochemistry and *in situ* hybridization

Ann Efthimiadis and Qutayba Hamid

Sputum cytology has been an important part of assessment for tuberculosis and cancer for decades, but it is only over the past 15 years that this method has been used extensively for the study of non-infectious inflammatory lung diseases. Initially, sputum was examined for the presence of eosinophils on stained smears. However, assessment of smears is difficult and counts unreliable

because of the inhomogeneous distribution of cells entrapped in mucus (Figure 2.1). The introduction of reducing agents such as dithiothreitol (DTT) or dithioerythritol (DTE) to disperse mucus and release trapped cells has made it possible to prepare homogeneous cell suspensions and an 'imprint' of these cells using a cytocentrifuge to generate cytospins with a monolayer of cells. This

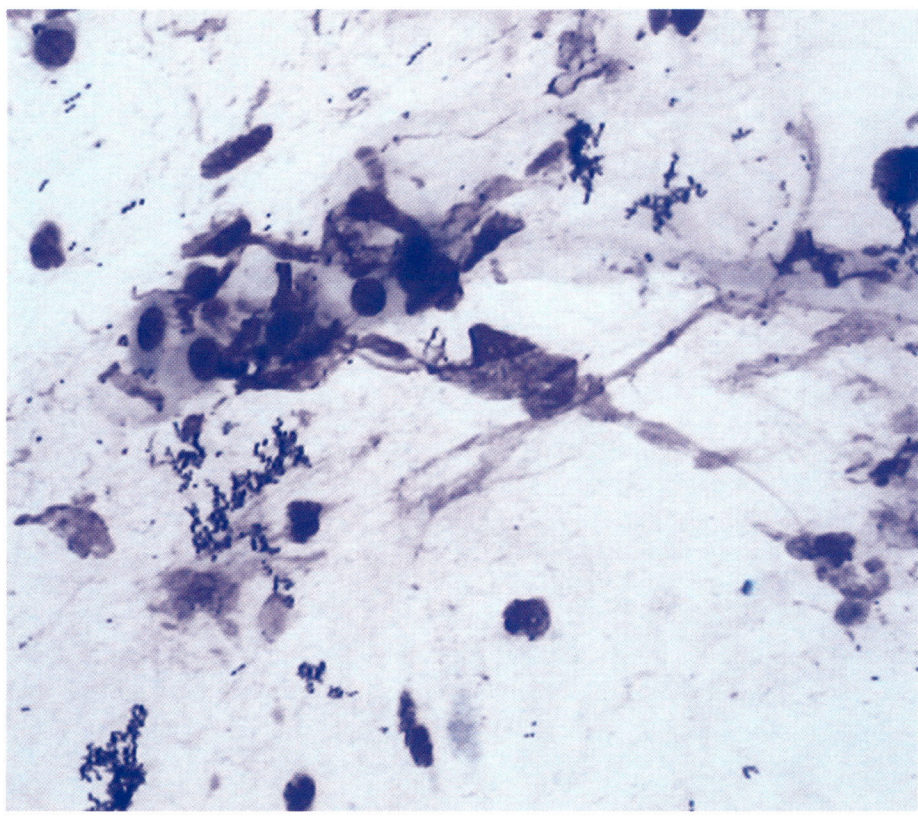

Figure 2.1 Typical smear of induced sputum. Various cell types are seen entrapped in mucus. Wright's stain, ×400 magnification

SPUTUM PROCESSING METHODS

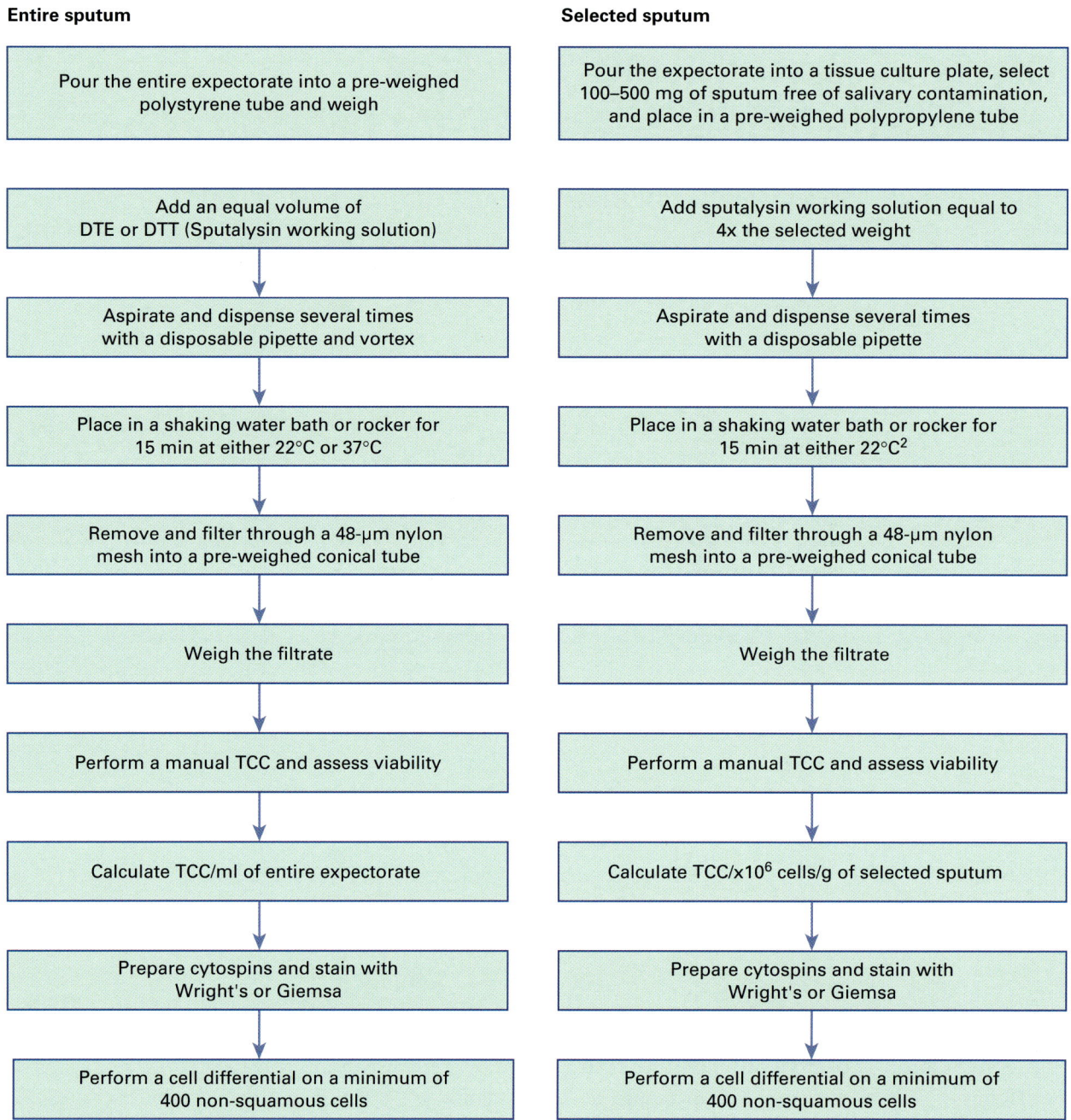

Figure 2.2 Two methods used for processing sputum: one method involves selection of viscous components and the other uses the entire expectorate. Reproduced with permission from Efthimiadis A, Spanevello A, Djukanovic R, *et al*. Methods of sputum processing for cell counts, immunocytochemistry and in situ hybridization. *Eur Respir J* 2002;37(Suppl):19s–23s

approach has made quantification and characterization of inflammatory cells in the airways reliable.

Currently there are two methods for processing sputum to obtain cell counts; both have been extensively evaluated and both yield accurate results provided they are performed in a standardized way. The first method requires selection of as little as 50 mg of the more dense portions of sputum with the aid of an inverted microscope, and the second involves processing the entire expectorate (sputum plus some saliva) (Figure 2.2).

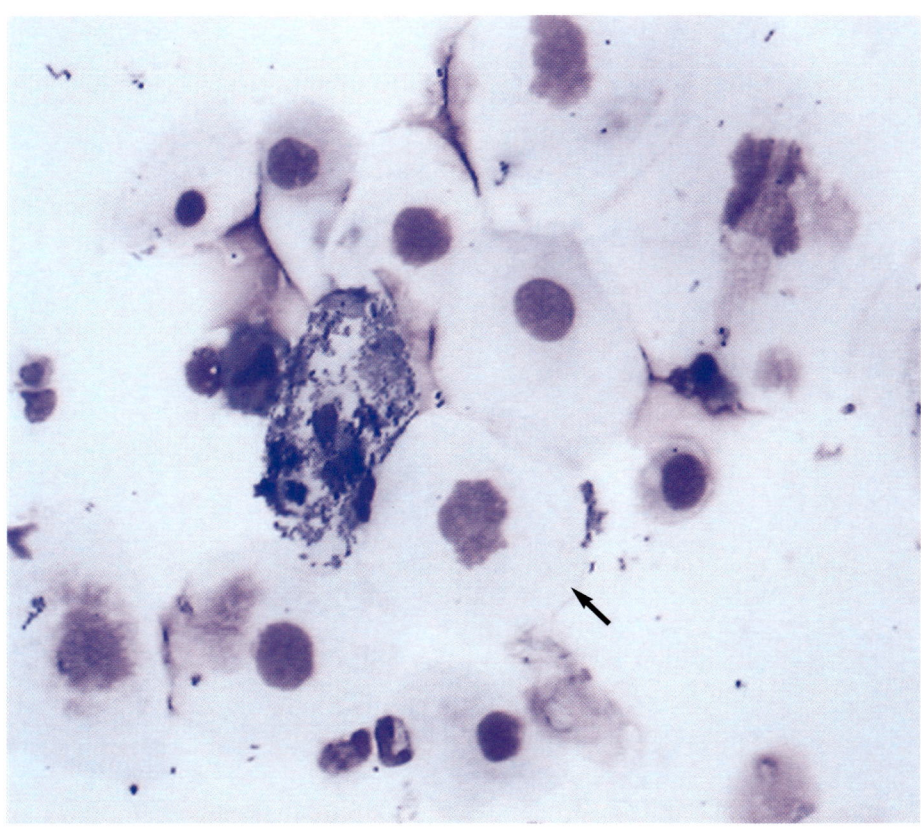

Figure 2.3 Typical cytospin preparation obtained by processing the entire expectorate. Samples processed in this way tend to contain more squamous cells (arrow), which may make it more difficult to count inflammatory cells. Wright's stain, x400 magnification

In both methods sputum is dispersed with reducing agents and filtered, after which the non-squamous total cell count and viability are determined in a hemocytometer. Cytospins are prepared and stained with Wright's or Giemsa stain for a differential cell count or stored for other staining techniques (Figures 2.3 and 2.4).

The advantages of using selected sputum rather than the entire expectorate are worthy of note. The total cell count (TCC), an important outcome to determine the severity of the inflammatory process, can be expressed per gram of lower airway secretions; quality cytospins can be produced with < 5% squamous cell contamination making characterization of cells easier and allowing additional specialized techniques to be performed such as polymerase chain reaction (PCR) and immunocytochemistry; and in addition, concentrations of chemicals in the fluid phase are unaffected by the confounding influence of variable dilution caused by saliva and can be more accurately corrected for dilution.

PROCESSING IN THE LABORATORY

Processing of sputum should be conducted as soon as possible and preferably within 2 h of expectoration. If there is any delay in processing, the sample can be stored in a refrigerator at 4ÐC for up to 8 h without affecting cell counts. After mixing the sputum with DTT or DTE to break up the mucus, the suspension is carefully filtered through a 48-mm nylon mesh to remove unwanted excess mucus and debris. Although there is a slight fall in the TCC after filtration, the relative proportions of cells in the differential cell count (DCC) are unaffected. Filtration is important and needs to be carefully performed to avoid excess mucus and debris in the filtrate. Mucus is known to interfere with staining intensity, producing dark staining that can make cell identification difficult and often impossible. Unwanted debris often contributes to non-specific antibody binding in immunocytochemical procedures and can have

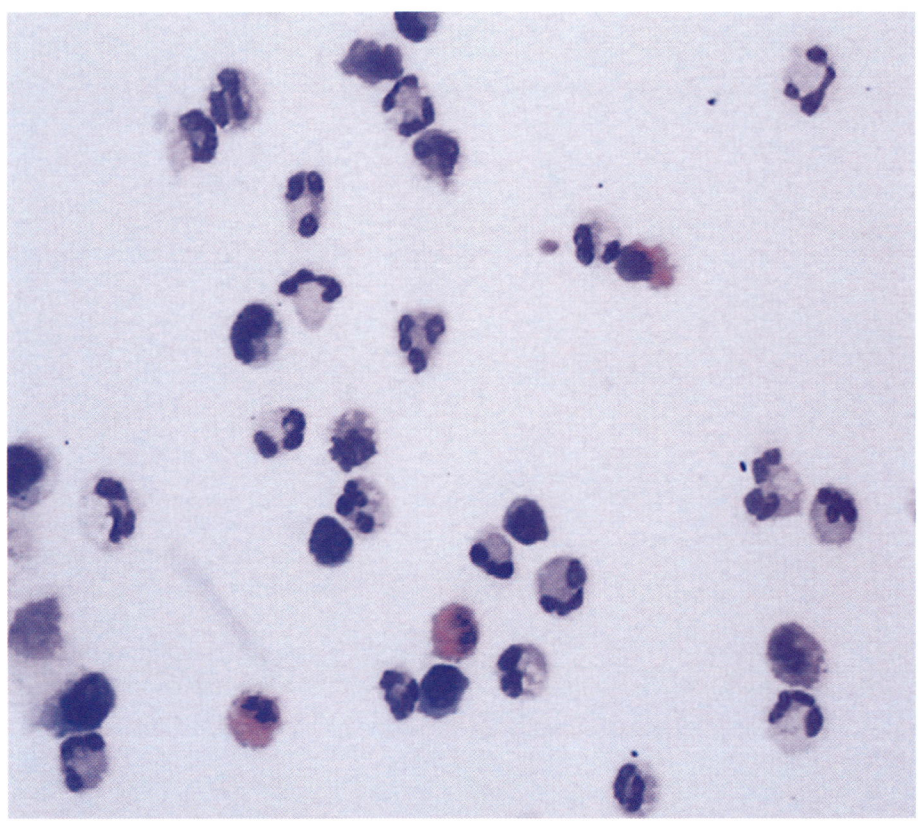

Figure 2.4 Typical cytospin preparation obtained by processing selected portions that are relatively free of salivary contamination. Wright's stain, ×400 magnification

an adverse effect on automated instruments by clogging their apertures. The manual TCC and viability are performed simultaneously with a Trypan blue stain and the remaining cell suspension is used for preparation of cytospins for routine staining with Wright's or Giemsa stains, identification of intracellular and surface cell markers by flow cytometry, or preparation of cytospins for immunocytochemistry and *in situ* hybridization.

EVALUATION AND DOCUMENTATION

The expectorated sample is macroscopically examined for color. The manual TCC/ml of selected sputum and percentage of viable cells are calculated. A minimum of 400 non-squamous cells is counted to determine the proportion of macrophages, neutrophils, eosinophils, lymphocytes and bronchial epithelial cells. The percentage of squamous cells is reported separately. The sputum report form should include the following information: patient characteristics; sputum induction spirometry data; the time of sample collection and processing; the color of sputum; the TCC and viability; the differential cell count; other comments worthy of note including presence of free eosinophil granules, apoptotic eosinophils and smokers inclusions in macrophages; and the signature of the technologist (Figure 2.5).

QUALITY CONTROL

Every laboratory requires an adequate quality control program, which forms an integral component of standard operating procedure (SOP) protocols. This is particularly important when slide readings are used to monitor patient treatment. Incorrect results can lead to incorrect diagnosis, treatment and research direction. First and foremost, it is mandatory that staff are both qualified and fully trained. Monthly quality control should include internal slide reading as well as equipment calibration; unless a certain standard of performance is fulfilled, 'normal values' have no significance.

Date of test:

Type   ☐ Spontaneous Sputum   ☐ Induced Sputum   ☐ Blown Nasal

Inhaled steroid dose   Prednisone dose _____ mg / day

Pre Induction Spirometry		Sputum Induction Spirometry	
Predicted FEV_1 / VC	L	FEV_1 post 0.9% normal saline	L
Baseline FEV_1 / VC	L	FEV_1 post % hypertonic / saline	L
Post Ventolin FEV_1 / VC	L	FEV_1 post % hypertonic / saline	L
Observer		FEV_1 post % hypertonic / saline	L
Collection time Processing time		Post Induction / bronchodilator FEV_1	L

Appearance   ☐ Mucoid   ☐ Muco-Purulent (in-between)   ☐ Purulent

Colour   ☐ Colourless   ☐ White   ☐ Grey   ☐ Green   ☐ Yellow   ☐ Brown   ☐ Red

	Normal Values – * 90th percentile
Viability _____ %	
Squamous cell contamination _____ % (not recorded if less than 20%)	> 40
Total cell count (of selected sputum) _____ $\times 10^6$ cells / mL	< 9.7*
Differential cell count	—
Neutrophils _____ %	< 64.4*
Eosinophils _____ %; of which _____ % are apoptotic	< 2.0
Macrophages _____ %	< 86.1*
Lymphocytes _____ %	< 2.6*
Bronchial epithelial _____ %	< 4.4*
Nasal epithelial _____ % Observer:	—

Other observations				Special stains	
Eosinophil free granules	few	moderate	many	Lipid / Oil Red-O Index	< 7
Macrophage smoker inclusions	few	moderate	many	% of macrophages hemosiderin +ve	> 2

Interpretation:

Figure 2.5 Example of a report on induced sputum, documenting changes in lung function during the induction caused by inhalation of saline and differential cell counts

Table 2.1 Differential cell counts in induced sputum from normal adults. Reproduced with permission from Belda J, Leigh R, Parameswaran K, *et al*. Induced sputum cell counts in healthy adults. *Am J Respir Crit Care Med* 2000;161:475–8

Sputum cell counts	Median (%)	IQR (%)	Percentiles (%)	
			10th	90th
Macrophages (%)	60.8	28.9	33.0	86.1
Neutrophils (%)	36.7	29.5	11.0	64.4
Eosinophils (%)	0.00	0.30	0.00	1.10
Lymphocytes (%)	0.50	1.80	0.01	2.60
Bronchial epithelial cells (%)	0.30	1.30	0.00	4.40

IQR, interquartile range

NORMAL FINDINGS

Non-smokers

In healthy subjects, cell counts show that macrophages and neutrophils predominate and that there are few eosinophils and lymphocytes (Table 2.1 and Figure 2.6).

Smokers

In otherwise healthy smokers, the total cell count is elevated and macrophages display blue/black, round/oval cytoplasmic inclusions, often referred to as 'smokers inclusions' (Figure 2.7).

CELL FINDINGS RELEVANT TO DISEASE

Abnormalities in DCC emphasise the occurrence of different types of inflammation and their different causes. The cellular features of the major conditions are discussed in this chapter; further details of results obtained in studies of the various diseases are given in other chapters.

An increased proportion of sputum eosinophils (more than 3%) is termed eosinophilic bronchitis (Figure 2.8). Eosinophilic bronchitis can be present in asthma and other airway diseases. In asthma, an influx of eosinophils into the airways, as identified in sputum, can be induced by inhaled allergen or chemical sensitizers in sensitized subjects. Similar changes can be seen following

reduction in steroid treatment in asthmatics treated with inhaled corticosteroids. Serial measurements during periods at work and off work can be used in the investigation of occupational asthma. However, eosinophilic bronchitis is a characteristic but not exclusive feature of asthmatic airway inflammation. Eosinophilic bronchitis can occur in the absence of variable airflow limitation and airway hyperresponsiveness in patients with chronic cough. Eosinophilic bronchitis can also be present in atopic cough without asthma, allergic rhinitis, and in a proportion of patients with chronic obstructive pulmonary disease (COPD). It is also important to note that symptoms and physiological abnormalities of asthma can be present without evidence of eosinophilic bronchitis and this may be associated with sputum neutrophilia. Several causes of sputum neutrophilia are known, including cigarette smoking, environmental pollutants such as ozone and diesel exhaust, chemical sensitizers, endotoxin and bacterial and viral infections (Figure 2.9).

Macrophages are the predominant cell type in healthy subjects and, in comparison with granulocytes, they have a longer life span and thus reside in the airways much longer. In conditions such as gastroesophageal reflux and left ventricular heart failure there is evidence of an accumulation of lipid-laden macrophages and hemosiderin-laden macrophages in sputum; these can be identified

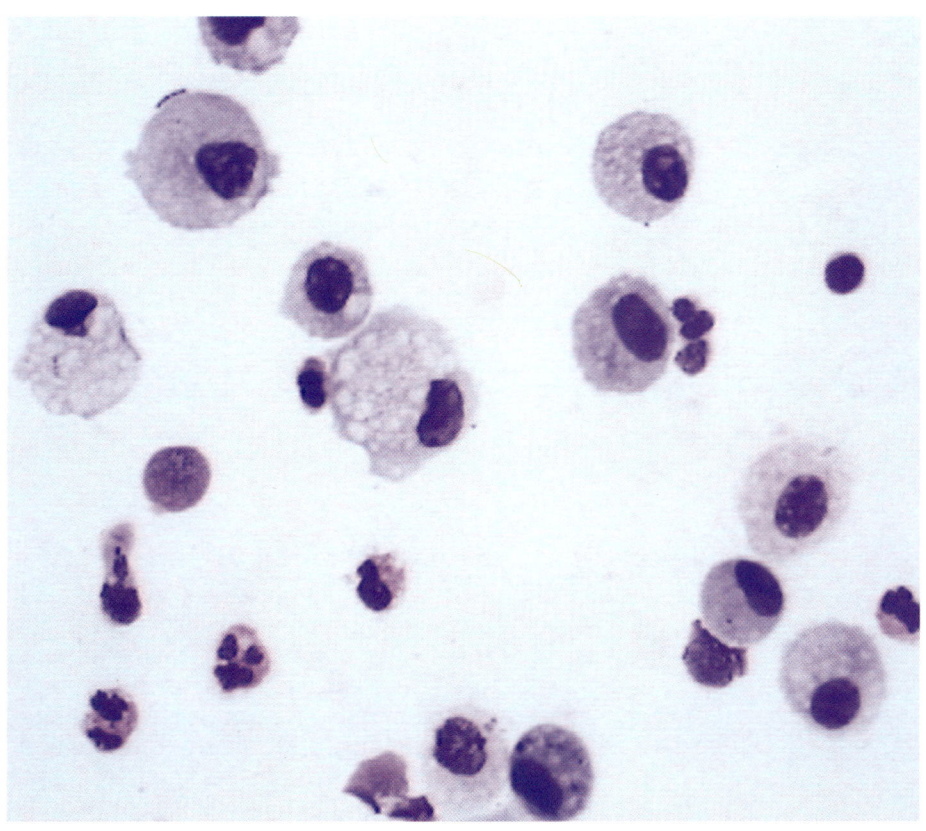

Figure 2.6 Induced sputum from a healthy non-smoking subject. Most of the cells are macrophages and neutrophils. Wright's stain, ×400 magnification

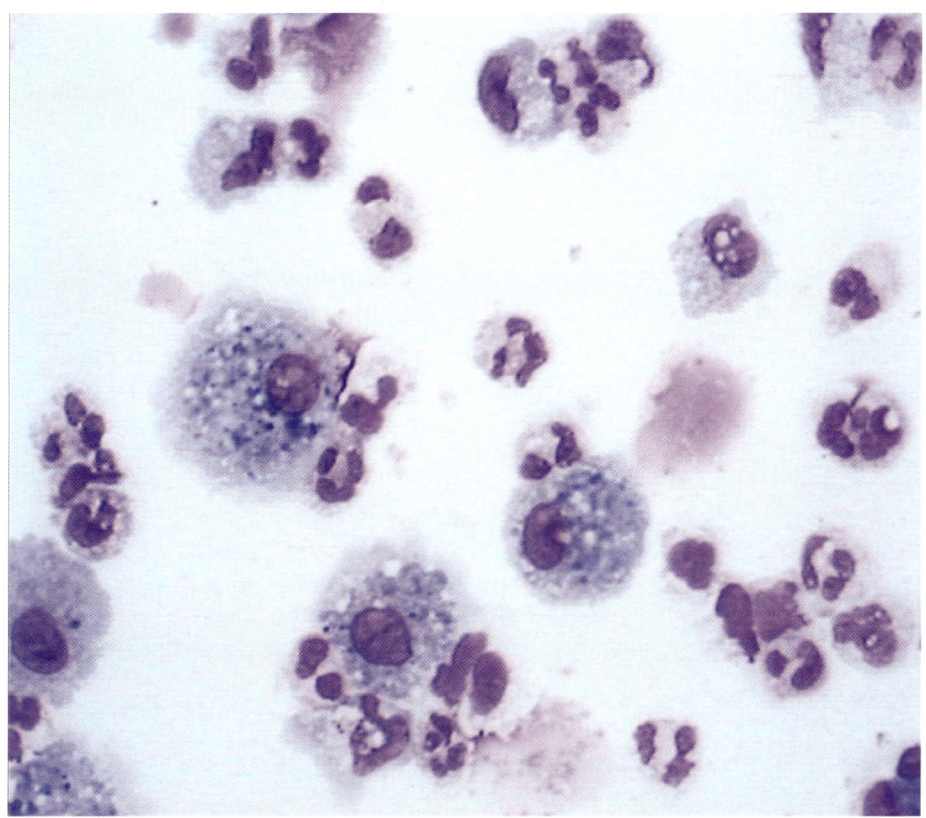

Figure 2.7 Induced sputum from a healthy smoker. The macrophages display 'smokers inclusions'. Wright's stain, ×400 magnification

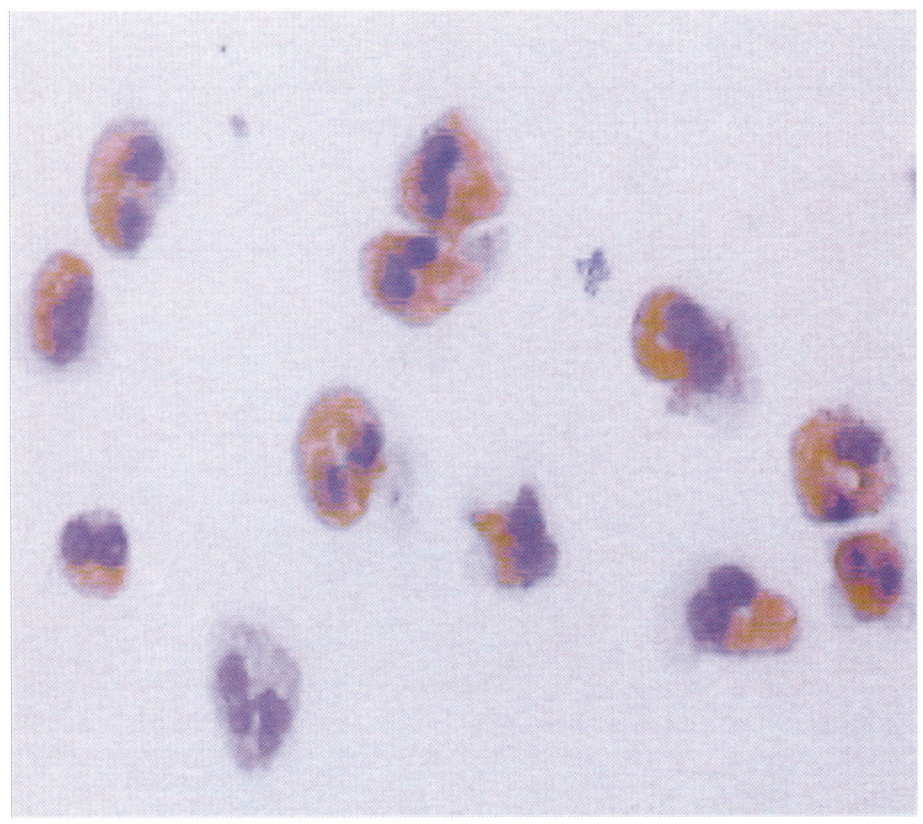

Figure 2.8 Induced sputum from a subject with eosinophilic asthma. Most of the cells are intact eosinophils with a bi-lobed nucleus, a feature of eosinophilic bronchitis. Wright's stain, ×540 magnification

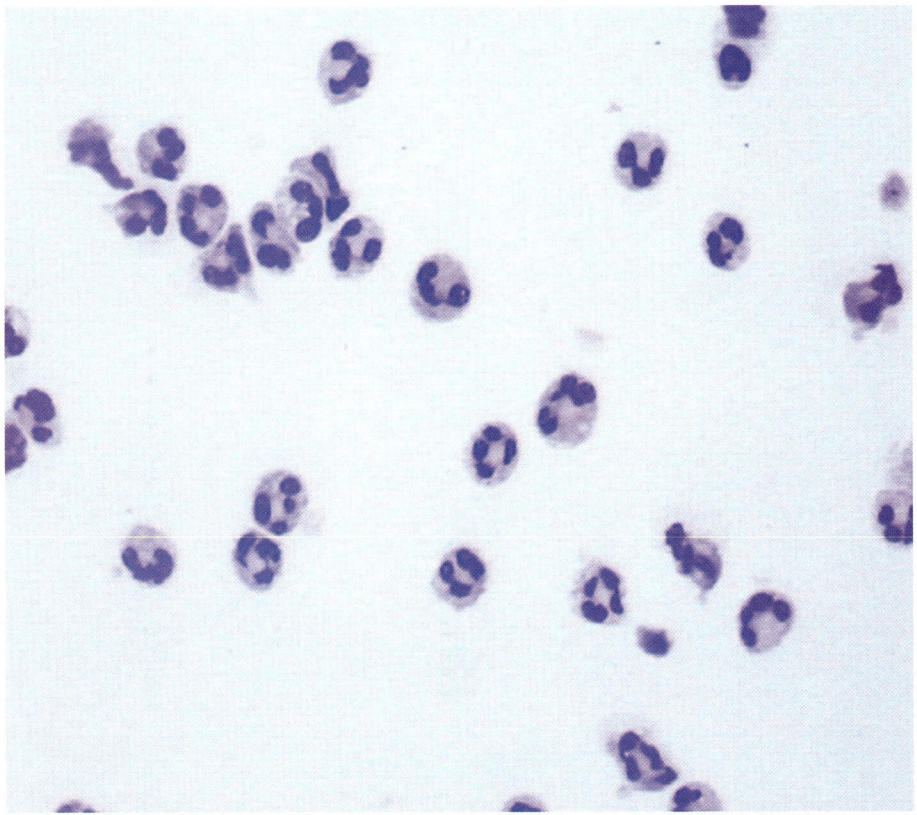

Figure 2.9 Induced sputum from a subject with neutrophilic asthma. Most cells are neutrophils, with some macrophages and occasional lymphocytes. Wright's stain, ×400 magnification

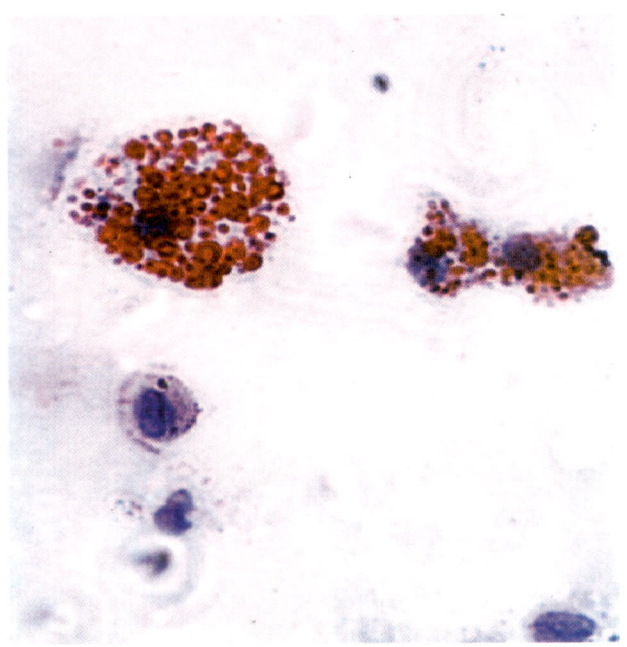

Figure 2.10 Induced sputum from a subject with gastroesophageal reflux. Droplet-shaped fat inclusions are seen in alveolar macrophages. A lipid index can be obtained by assessing 100 macrophages and grading the lipid content. A lipid index of 7 has been shown to have a good predictive value for gastroesophageal reflux. Oil Red-O stain, ×400 magnification

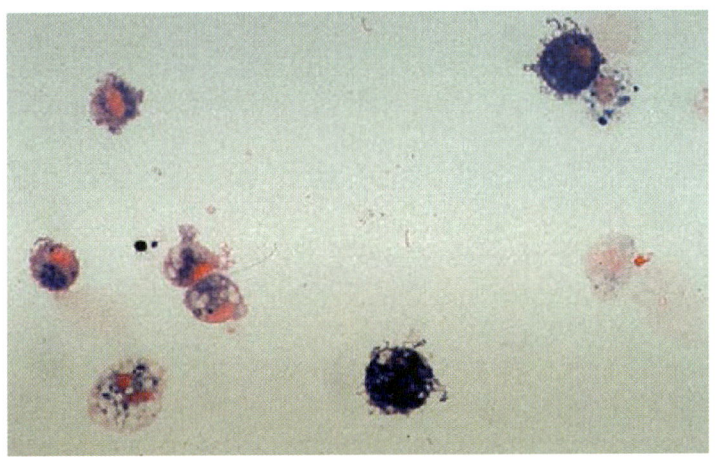

Figure 2.11 Induced sputum from a subject with left ventricular dysfunction. Alveolar macrophages that are rich in hemosiderin display blue to black staining. Prussian blue stain, ×400 magnification

and enumerated using Oil Red-O and Prussian blue staining respectively (Figures 2.10 and 2.11).

The demonstration of lipid-laden macrophages has been suggested as a useful diagnostic test for micro-aspiration associated with gastroesophageal reflux. A lipid index can be calculated by observing macrophages in sputum cytospins; this is a composite score derived after assessing 100 macrophages and grading their intracellular lipid content from 0 (no lipid droplets) to 4 (many intracellular oil red-O-stained lipid droplets obscuring the nucleus). The grades for each cell are added to give a lipid index that can range from 0 to 400. A lipid index greater than 7 has been found to have a sensitivity of 90% and a specificity of 89%, with positive and negative predictive values of 95 and 80%, respectively.

The presence of macrophages containing hemosiderin may be a marker of left ventricular dysfunction in breathless patients suspected of having cardiac or respiratory disease. Counts of hemosiderin-containing macrophages above 2%

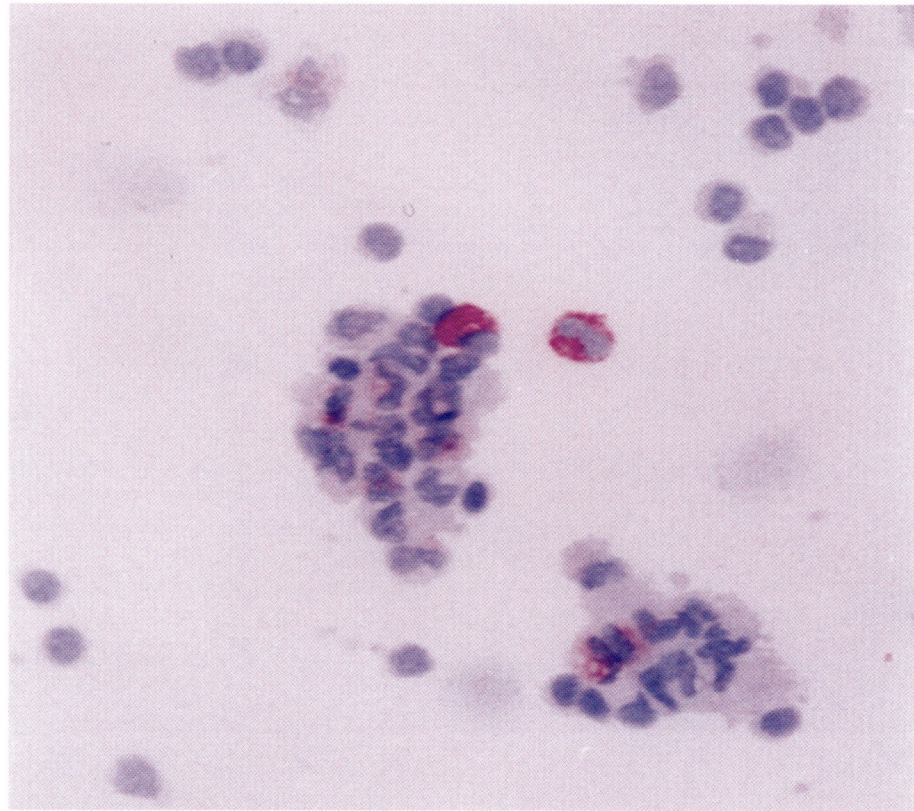

Figure 2.12 Immunocyto-chemical staining of a cytospin preparation of sputum, showing a number of immunoreactive eosinophils following staining with antibody against major basic protein using the APAAP technique. The cytoplasm of positive cells stains red. Magnification ×400

are accurate for the presence of left ventricular dysfunction confirmed by 2D electrocardiographic criteria, with a positive predictive value of 96%.

IMMUNOCYTOCHEMISTRY AND *IN SITU* HYBRIDIZATION OF SPUTUM SPECIMENS

Immunocytochemistry can be performed on cytospin preparations from sputum specimens. The cytospin preparation must be fixed with acetone/ methanol for 7 min prior to staining. The recommended method is the alkaline phosphatase anti-alkaline phosphatase (APAAP) technique (Figure 2.12). Double staining using antibodies to two different cell markers, such as CD3 for T cells and CD68 for macrophages, can also be applied, using a combination of staining methods (Figure 2.13). The APAAP technique involves incubation of cytospin samples with monoclonal antibodies and the development of antibody–antigen complex. Polyclonal antibodies and other

methods such as the peroxidase anti-peroxidase (PAP) and the avidin biotin conjugated (ABC) immunocytochemical techniques can be used. The immunofluorescence technique is not recommended for sputum samples, as it is difficult to differentiate between specific signal and non-specific signal due to auto-fluorescence of cells in sputum. In order to minimize background staining, action should be taken to reduce squamous cell contamination and the presence of mucus. Negative controls should always be used.

A number of laboratories have used immunocytochemistry in their research and have reported successful staining for different cell types, employing antibodies that have been previously used in bronchial biopsies (Figure 2.12). It appears that there is some discrepancy between the number of cells identified by conventional staining and the number of cells identified by immunocytochemistry.

A study that compared Wright's staining and immunocytochemistry showed good agreement for

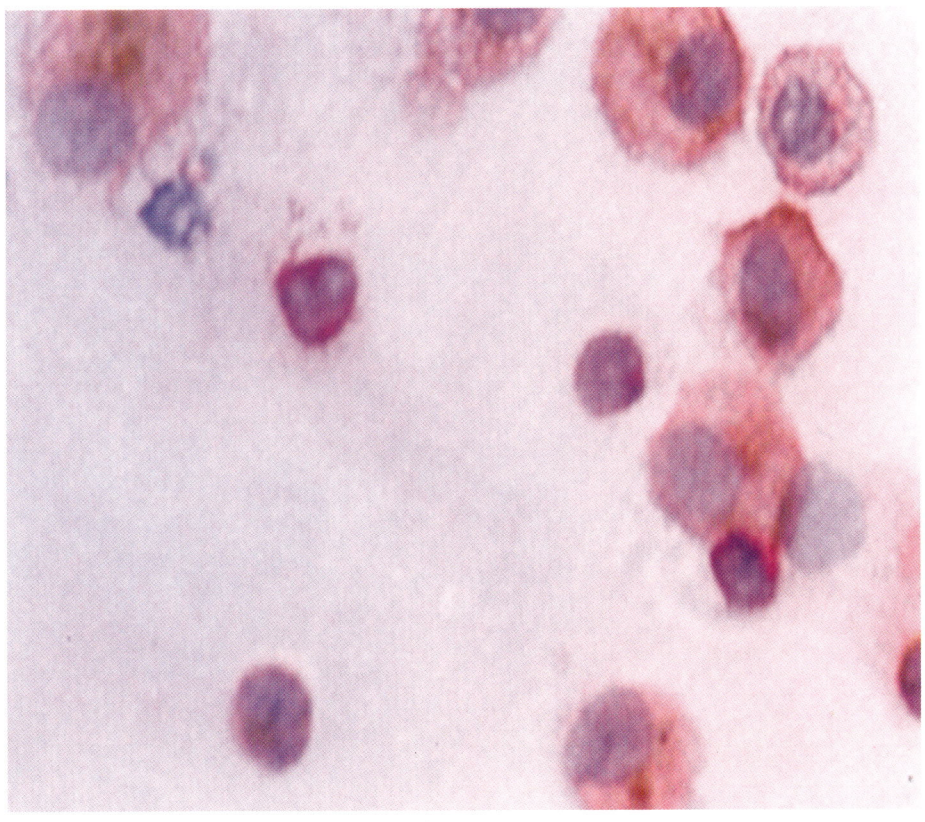

Figure 2.13 Double immuno-cytochemistry using the APPAP (red cells) and PAP (brown cells) techniques performed on a cytospin preparation using antibodies to CD3 and CD68; these identify lymphocytes and macrophages, respectively. Magnification ×400

eosinophils, less agreement for neutrophils and macrophages, and poor agreement for lymphocytes and epithelial cells (Figure 2.14). The discrepancy between lymphocyte and epithelial cell counts can be explained in part by the fact that the recognition of cell morphology is more subjective with Wright staining than with immunocytochemistry, and it has to be stated that for most studies standard stains such as Wright's and Giemsa are perfectly adequate for eosinophils, macrophages and neutrophils.

Cytospin preparations are also a suitable means of detecting cell-associated immunoreactivity of cytokines and chemokines. Studies have shown expression of eotaxin, interleukin (IL)-16 and MCP-4 in sputum preparations from asthmatics (Figure 2.15). Cytokine receptor and transcription factor immunoreactivities can also be detected in sputum in asthma, and these are significantly higher than in control subjects. Another example of the usefulness of this method is the immunoreactivity for cyclo-oxygenase (COX) enzymes 1 and 2, which can be detected in sputum from asthmatics and COPD subjects.

Many cytokines are detected with difficulty by immunocytochemistry and, therefore, need to be identified at the mRNA level. *In situ* hybridization is a technique that allows the detection of intracellular mRNA using radioactive or non-radioactive antisense probes (Figure 2.16). Preparation of sputum samples for this procedure requires special care, without which interpretation can be very difficult (Figure 2.17). Sputum samples should be treated with special consideration to avoid RNA degradation or contamination with RNases. Radioactive probes can attach non-specifically to mucus as well as inflammatory cells. It is therefore essential to use high stringency washing and to use a sense probe as a control. Increased expression of cytokine mRNA, including that for IL-4 and IL-5, has been reported in asthmatics when compared with control subjects using *in situ* hybridization.

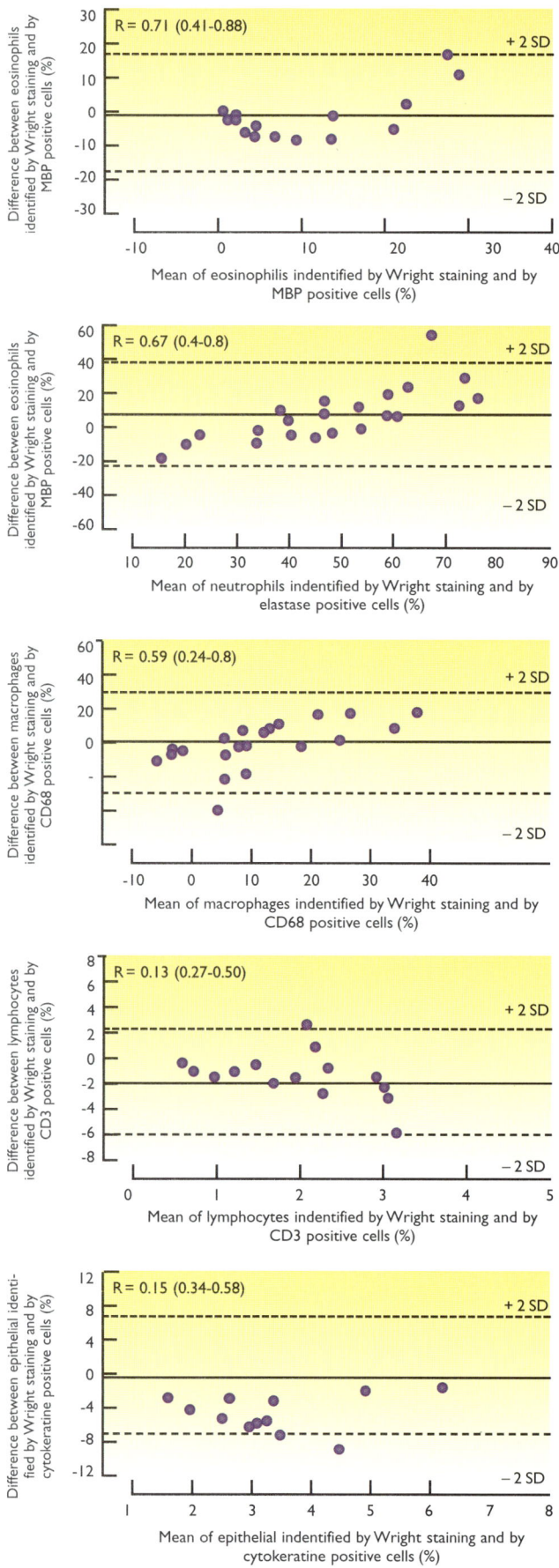

Figure 2.14 Agreement (%) between sputum neutrophil, macrophage, eosinophil, lymphocyte, and epithelial cell counts after Wright's staining and after immunocytochemistry. R is the intraclass correlation coefficient, and the shaded area represents ± 2 standard deviations (SD) of the mean of the differences. Reproduced with permission from Lemiere C, Taha R, Olivenstein R, Hamid Q. Comparison of cellular composition of induced sputum analyzed with Wright staining and immunocytochemistry. *J Allergy Clin Immunol* 2001;108:521–3

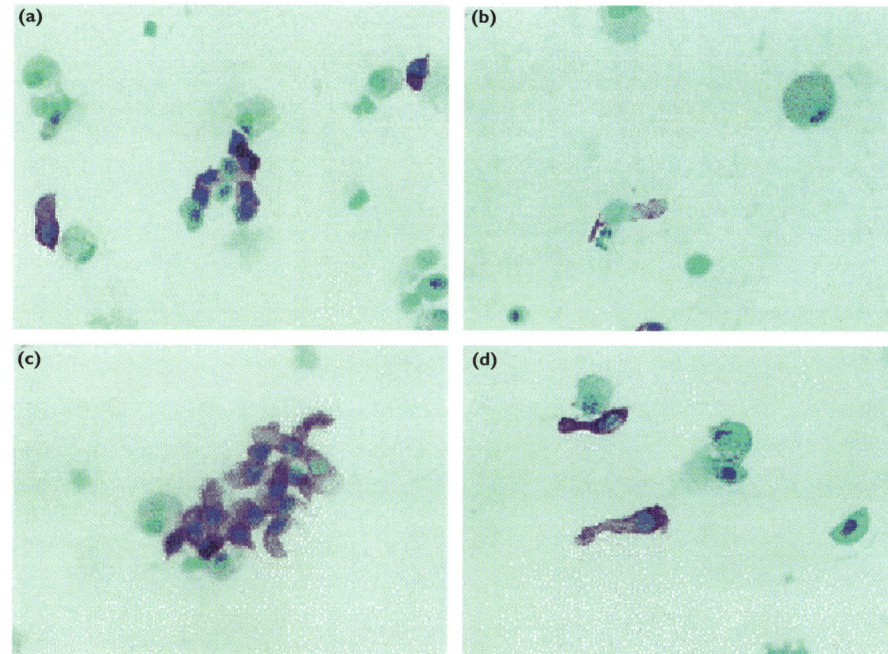

Figure 2.15 Representative examples of immunocytochemistry of epithelial cells expressing eotaxin and MCP-4 immunoreactivity in induced sputum from asthmatic patients and normal control subjects (haematoxylin, original ×200). Red cytoplasmic staining in top left (a) and top right (b) indicates eotaxin immunoreactivity in asthmatic patients and normal control subjects, respectively. Bottom left (c) and bottom right (d) show MCP-4 immunoreactivity in asthmatic patients and normal control subjects, respectively. Reproduced with permission from Taha RA, Laberge S, Hamid Q, Olivenstein R. Increased expression of the chemoattractant cytokines eotaxin, MCP-4 and IL-16 in induced sputum in asthmatics. *Chest* 2001;120:595–601

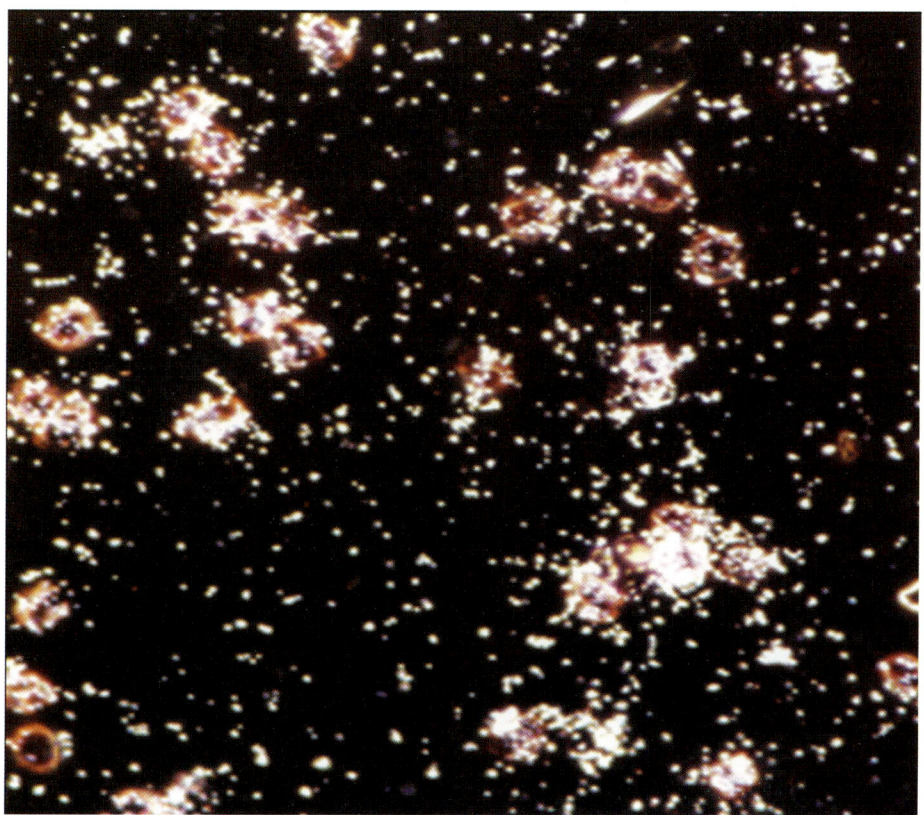

Figure 2.16 Positive staining for TNF-α mRNA in sputum cells identified by *in situ* hybridization using radioactive antisense cRNA probes and darkfield illumination. Positive cells are covered with white-silver grains. Magnification ×400

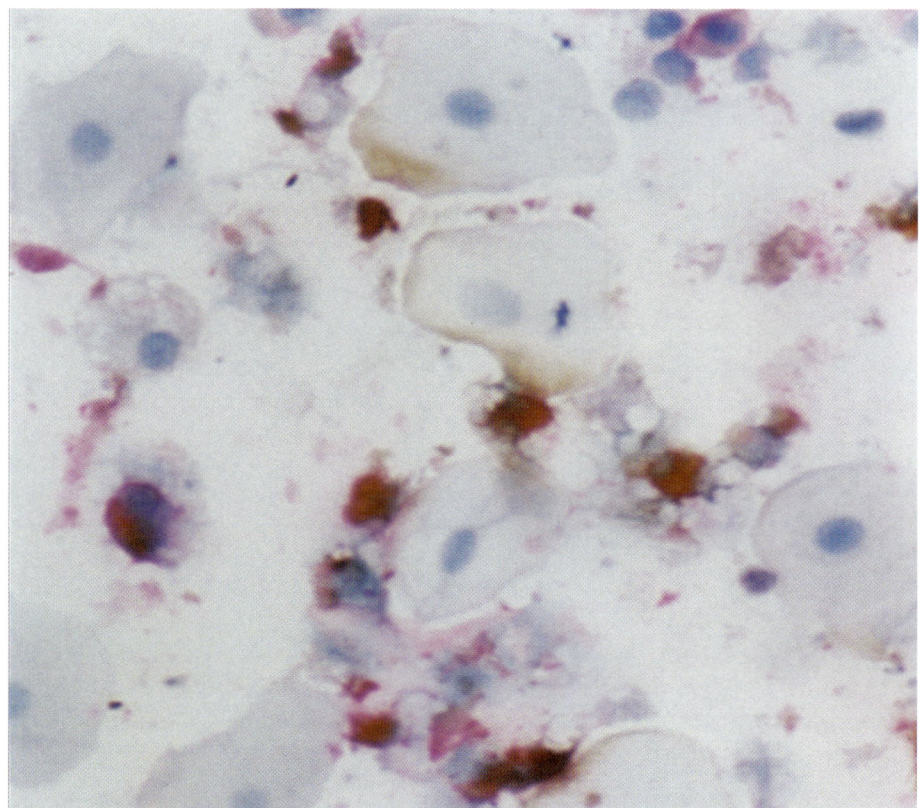

Figure 2.17 An example of an unacceptable sputum preparation, immunostained with antibodies against major basic protein employing the APAAP method. Note the high background due to presence of mucus and the large number of squamous cells. Such inadequate sputum preparations make immunocytochemistry very difficult to interpret and should not be used

In summary, sputum cytology is a central component of sputum analysis and one that must be included in any analysis of sputum, even if the main objectives of the study is to detect soluble mediators. Correct interpretation requires careful quality control of all stages of sputum processing.

The application of techniques such as immuno-histochemistry and *in situ* hybridization, although not yet as widespread as that of basic staining, offers significant opportunities for further research and possibly diagnostics.

CHAPTER 3

Analysis of soluble mediators in the sputum fluid phase

Margaret Kelly and Janis Shute

Sputum contains viscous mucus that requires dispersion in order to release soluble mediators so that they can be measured. The large number of soluble mediators measured in sputum has been matched by the large variety of sputum processing techniques applied, even for the measurement of the same mediator; this and the use of different assay methods have made it difficult to compare measurements conducted in different centres. Despite this, measurements of soluble mediators are often included when sputum cellular indices are evaluated, and the results can be used to support hypotheses.

THE COMPOSITION OF SPUTUM

Sputum is a complex mixture of inflammatory cells, debris, exfoliated epithelial cells, water, ions, proteins and lipids, all held together within a mucus gel. The physical properties of mucus are determined by the presence of mucus glycoproteins (mucins) which consist of large peptide molecules forming a backbone upon which are attached numerous oligosaccharide side-chains. The large number of hydrophobic and hydrophilic sites present in these molecules allow multiple low-affinity bonds to occur with virtually any substance, accounting for the marked 'stickiness' of mucus. Thus, numerous cytokines, growth factors and other molecules may adhere to mucus and constitute what is referred to as the soluble sputum content.

The tertiary structure of mucus is maintained by numerous disulphide bonds which link glycoprotein subunits into extended mucin oligomers. Disruption of the disulphide bonds by strong reducing agents, such as dithiothreitol (DTT), allows dispersion of sputum and the release of soluble molecules.

COMPARISONS BETWEEN INDUCED SPUTUM AND BRONCHOALVEOLAR LAVAGE FLUID

Bronchoalveolar lavage (BAL) has been widely used to study lung inflammation in a variety of diseases including asthma and chronic obstructive pulmonary disease (COPD). Apart from the fact that sputum samples can be obtained more safely and easily, sputum has the potential advantage over BAL that soluble mediators are more concentrated. Figure 3.1 shows that levels of eosinophil cationic protein (ECP) and fibrinogen are higher in induced sputum than in BAL.

There are several factors that can affect the concentration of various mediators in sputum (Table 3.1), including the method of induction and the method of processing the expectorated sample.

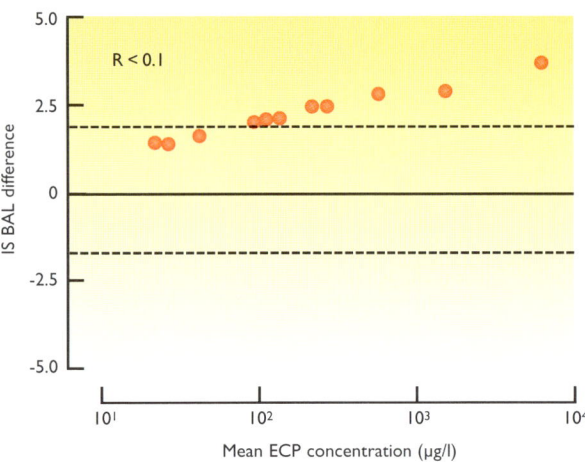
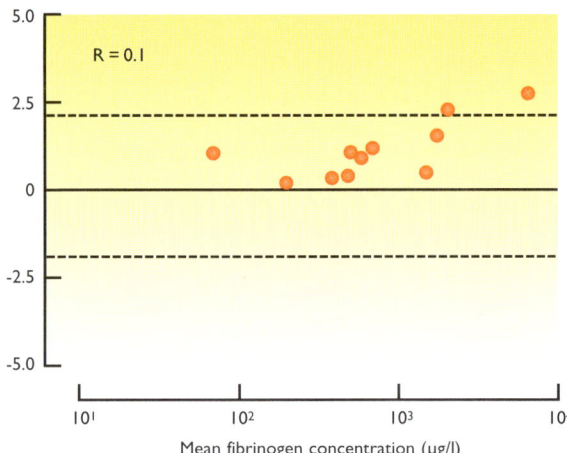

Figure 3.1 Comparisons of inflammatory indices in bronchoalveolar lavage (BAL) and sputum. The differences (log values) between induced sputum (IS) and BAL eosinophil cationic protein (ECP) and fibrinogen concentrations are plotted as a function of the mean of the two values. Intraclass correlation coefficients (R) are shown. Solid lines represent the mean difference; dashed lines represent ± 2 standard deviations (SD) of mean difference. In a Bland–Altman graph, if the variable is reproducible, it is expected that 95% of the differences between measures are less than 2 SD. Since all the data points are present above the line depicting the mean difference, this indicates that the levels in IS are consistently higher than those in BAL. Adapted with permission from Pizzichini E, Pizzichini MM, Kidney JC, *et al*. Induced sputum, bronchoalveolar lavage and blood from mild asthmatics: inflammatory cells, lymphocyte subsets and soluble markers compared. *Eur Respir J* 1998;11:828–34

Table 3.1 Factors that influence interpretation of mediator measurements in sputum

1. Method of induction:
   a. spontaneously expectorated sputum or induced by saline
   b. use of normal (isotonic) or hypertonic saline
   c. duration of induction
   d. volume of nebulized saline delivered
   e. interval between inductions
2. Preparation and processing of sputum
   a. sampling of selected or entire (unselected) sputum
   b. processing with reducing agents
   c. processing with saline
   d. temperature
   e. addition of protease inhibitors

FACTORS RELATED TO THE METHOD OF SPUTUM INDUCTION OF RELEVANCE TO MEDIATOR MEASUREMENT

Few individuals, except for those with purulent pulmonary conditions such as cystic fibrosis or COPD, can expectorate spontaneously. Nevertheless, in some cases, especially when safety is the issue, a sample has to suffice. This can have a major effect on mediator measurement, which must be considered when interpreting the data. The concentration of fibrinogen has been found to be higher in spontaneously expectorated sputum than in an induced sample (Figure 3.2). There is also a trend towards increased ECP in the spontaneous sample.

The duration of sputum induction also plays a role. It affects not only cell counts (as discussed in Chapter 1), but also the concentrations of mediators, with ECP, albumin and lactate dehydrogenase (LDH) decreasing with time (Figures 3.3 and 3.4). In addition, repeated sputum induction within 30 min results in reduced concentrations of ECP and interleukin (IL)-8, as shown in Figure 3.5. The reduction in the concentration of mediators with increased duration of induction may be related to dilution by the inhaled fluid. Standardization of the duration of induction is, therefore, strongly recommended when comparing mediators in different samples.

As well as the length of time of induction, the type of nebulizer is also important, with higher

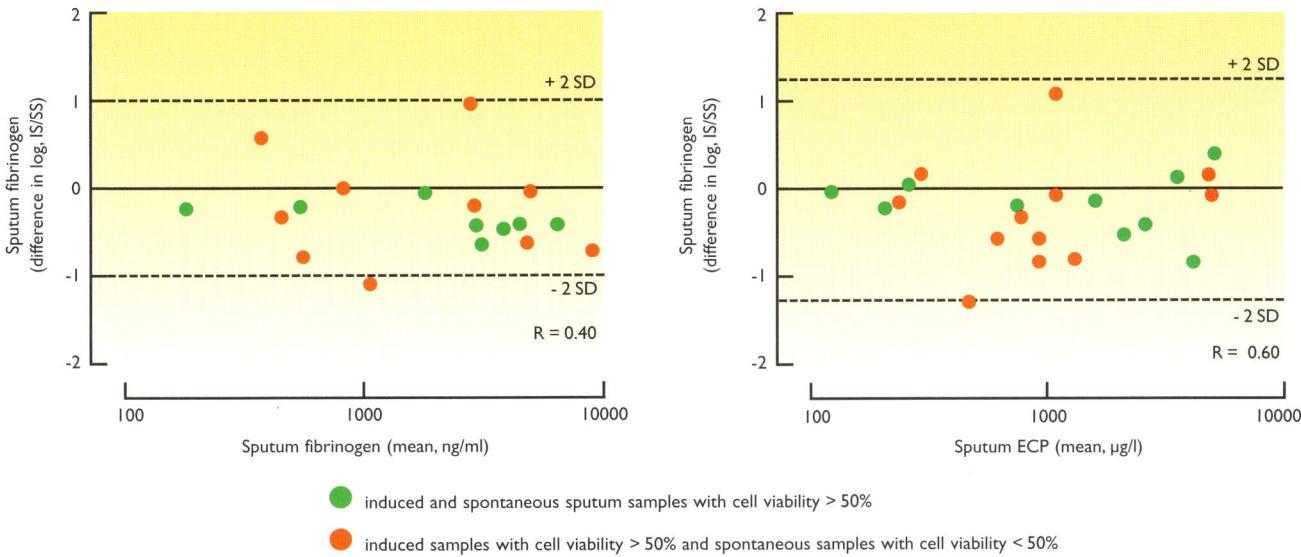

Figure 3.2 Comparisons of inflammatory indices in sputum obtained spontaneously or following induction. The difference (in logs) between induced (IS) and spontaneous (SS) sputum fibrinogen and eosinophil cationic protein (ECP) is plotted against the mean of the two values. R is the intraclass correlation coefficient and the dottted lines represent ± 2 standard deviations (SD) of the mean of the two differences. Adapted with permission from Pizzichini MM, Popov TA, Efthimiadis A, *et al*. Spontaneous and induced sputum to measure indices of airway inflammation in asthma. *Am J Respir Crit Care Med* 1996;154:866–9

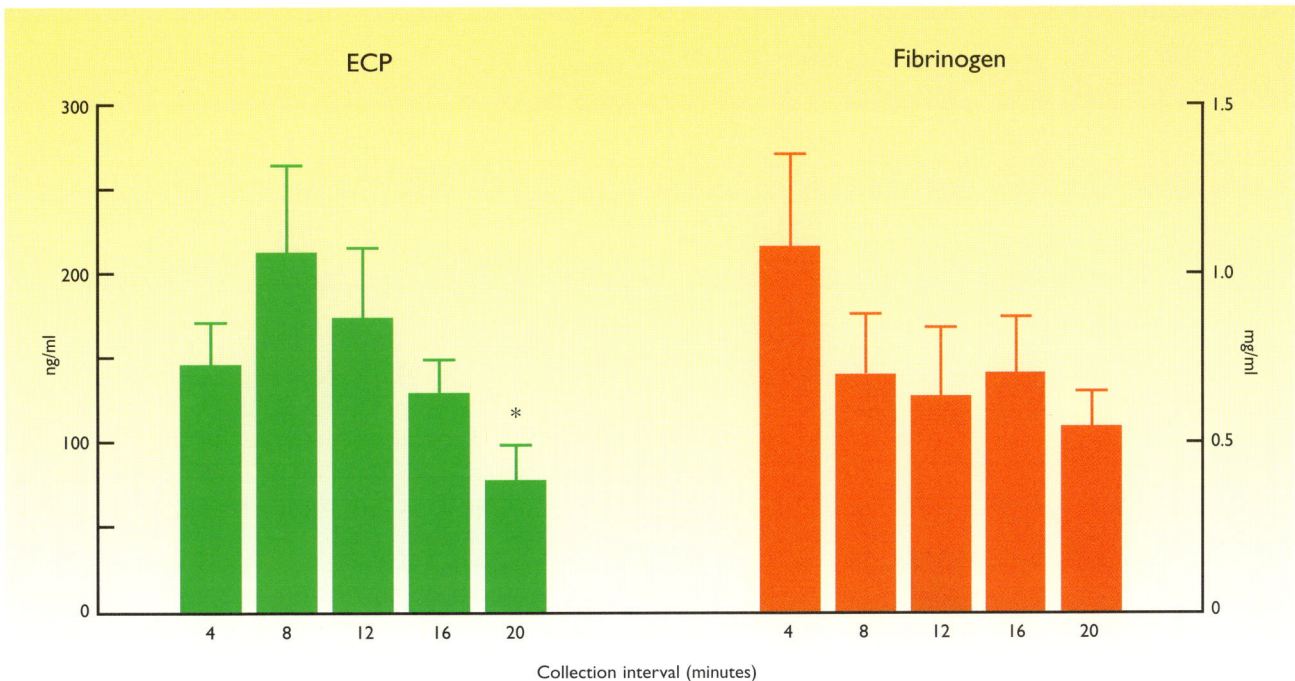

Figure 3.3 Changes in eosinophil cationic protein (ECP) and fibrinogen concentrations in induced sputum samples collected at five 4-min intervals during a 20-min sputum induction and analyzed separately. *Significantly different from the 0–4 min interval ($p < 0.05$). Reproduced with permission from Gershman NH, Liu H, Wong HH, *et al*. Fractional analysis of sequential induced sputum samples during sputum induction: evidence that different lung compartments are sampled at different time-points. *J Allergy Clin Immunol* 1999;104:322–8

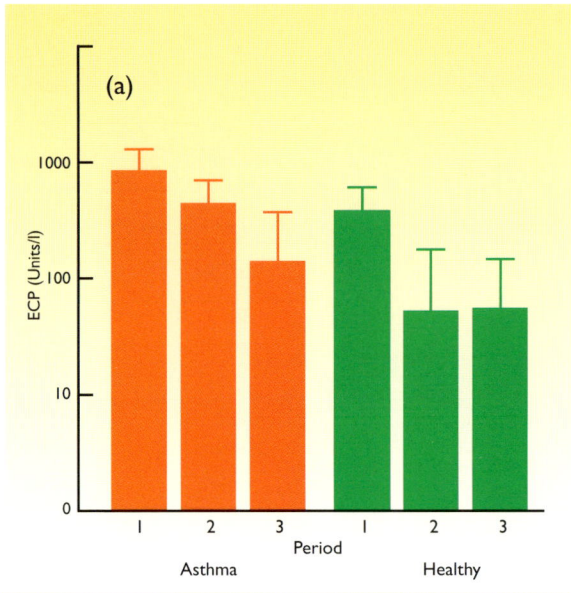

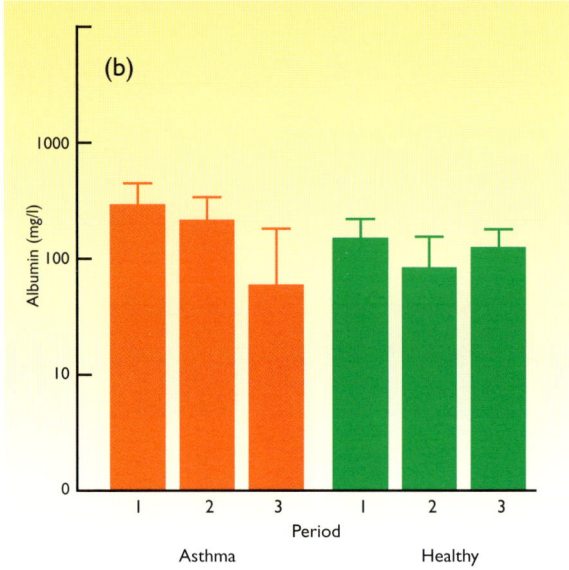

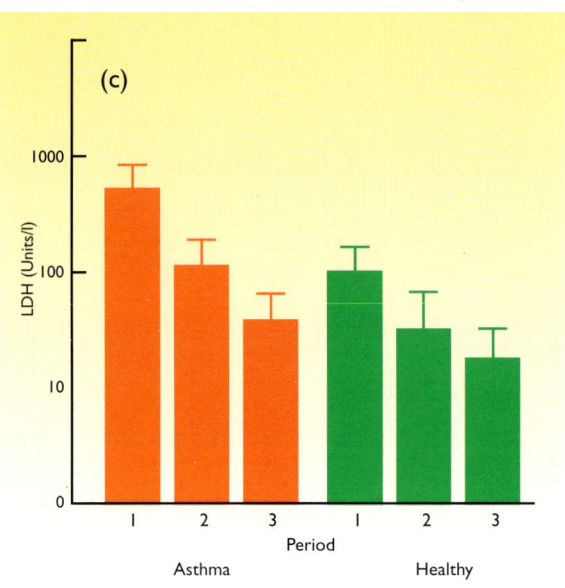

Figure 3.4 Changes in inflammatory indices in sputum in relation to time in asthmatic and healthy control donors. The mediator concentrations in three consecutive sputum samples obtained at 10-min intervals within one induction procedure are shown as geometric mean for (a) eosinophil cationic protein (ECP); (b) albumin; and (c) lactate dehydrogenase (LDH). The decline in the concentration of ECP was statistically significant in both healthy ($p < 0.001$) and asthmatic subjects ($p < 0.001$). The same was true for LDH ($p < 0.01$; $p < 0.05$). Changes in albumin levels were statistically significant only when subjects with asthma were analyzed separately ($p < 0.01$). Reproduced with permission from Holz O, Jorres RA, Koschyk S, *et al.* Changes in sputum composition during sputum induction in healthy and asthmatic subjects. *Clin Exp Allergy* 1998;28:284–92

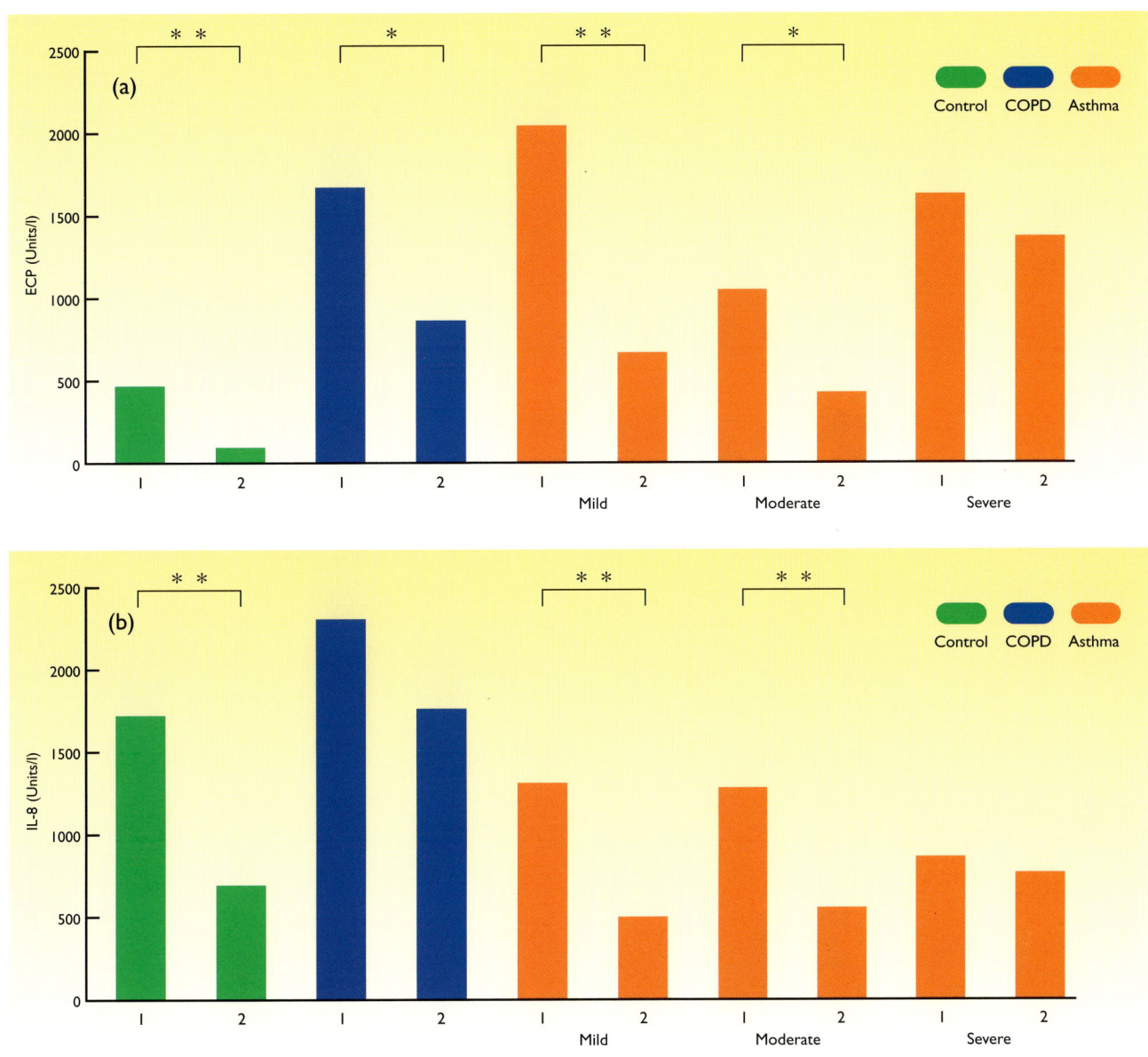

Figure 3.5 The effect of repeat (after 30 min) induction on sputum concentrations of eosinophil cationic protein (ECP) and interleukin (IL)-8. Geometric means of (a) ECP and (b) IL-8 concentrations in 10-min sputum inductions performed 30 min apart (1 and 2, respectively) in control subjects, subjects with COPD and subjects with mild, moderate and severe asthma. $*p < 0.05$; $**p < 0.01$ between first and second induction. Reproduced with permission from Richter K, Holz O, Jorres RA, *et al*. Sequentially induced sputum in patients with asthma or chronic obstructive pulmonary disease. *Eur Respir J* 1999;14:697–701

output nebulizers producing sputum with lower ECP and IL-8 concentrations (Figure 3.6).

The interval between sputum inductions is also a factor to consider. Levels of ECP are increased when sputum is induced within 24 h of a previous induction, indicating that the induction procedure itself causes a change in the composition of sputum detectable after 24 h (Figure 3.7).

Finally, given that induction with normal, as opposed to hypertonic, saline may be more suitable for some patients, it is important to understand whether this may influence the levels of mediators. Inhalation of isotonic saline solution, compared with hypertonic saline solution (4.5%), was shown in one study to result in comparable total and differential sputum cell counts, and levels

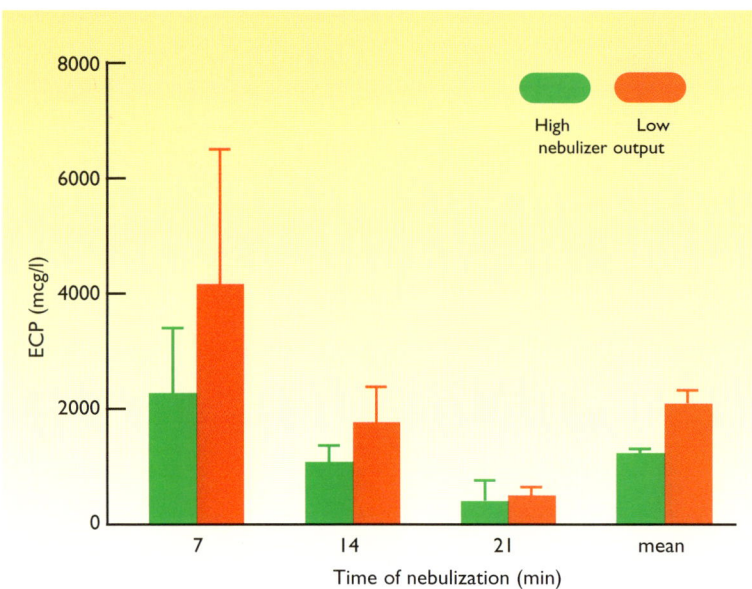

Figure 3.6 The combination of the effects of nebulizer output and the time of nebulization on selected sputum outcomes. Dispersion bars are standard errors. The higher output nebulizer resulted in significantly lower eosinophil cationic protein (ECP) ($p < 0.01$) and interleukin (IL)-8 ($p < 0.01$) levels than the lower output nebulizer, independent of duration of nebulization. The duration of nebulization also had a significant effect on the levels of ECP ($p < 0.01$). The results for IL-8 are not shown but IL-8 levels were similarly affected by the type of nebulizer and duration of nebulization. Reproduced with permission from Belda J, Hussack P, Dolovich M, *et al*. Sputum induction: effect of nebulizer output and inhalation time on cell counts and fluid-phase measures. *Clin Exp Allergy* 2001;31:1740–4

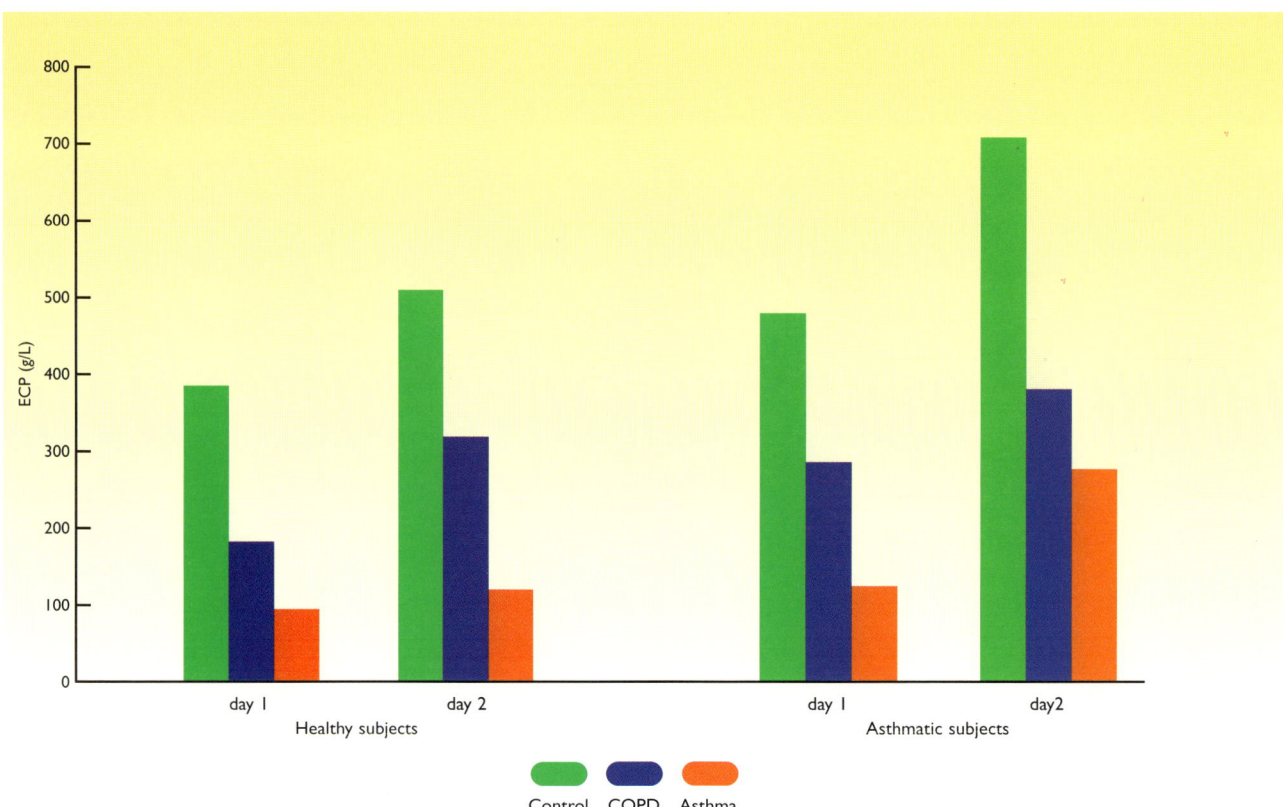

Figure 3.7 Consecutive sputum inhalation periods of 10 min each (three groups of 10 min shown) performed at the same time on two consecutive days in healthy and asthmatic subjects. Values are given as geometric means. Pooled eosinophil cationic protein (ECP) concentrations were significantly increased on day 2 compared with day 1 ($p < 0.05$). Reproduced with permission from Holz O, Richter K, Jorres RA, *et al*. Changes in sputum composition between two inductions performed on consecutive days. *Thorax* 1998;53:83–6

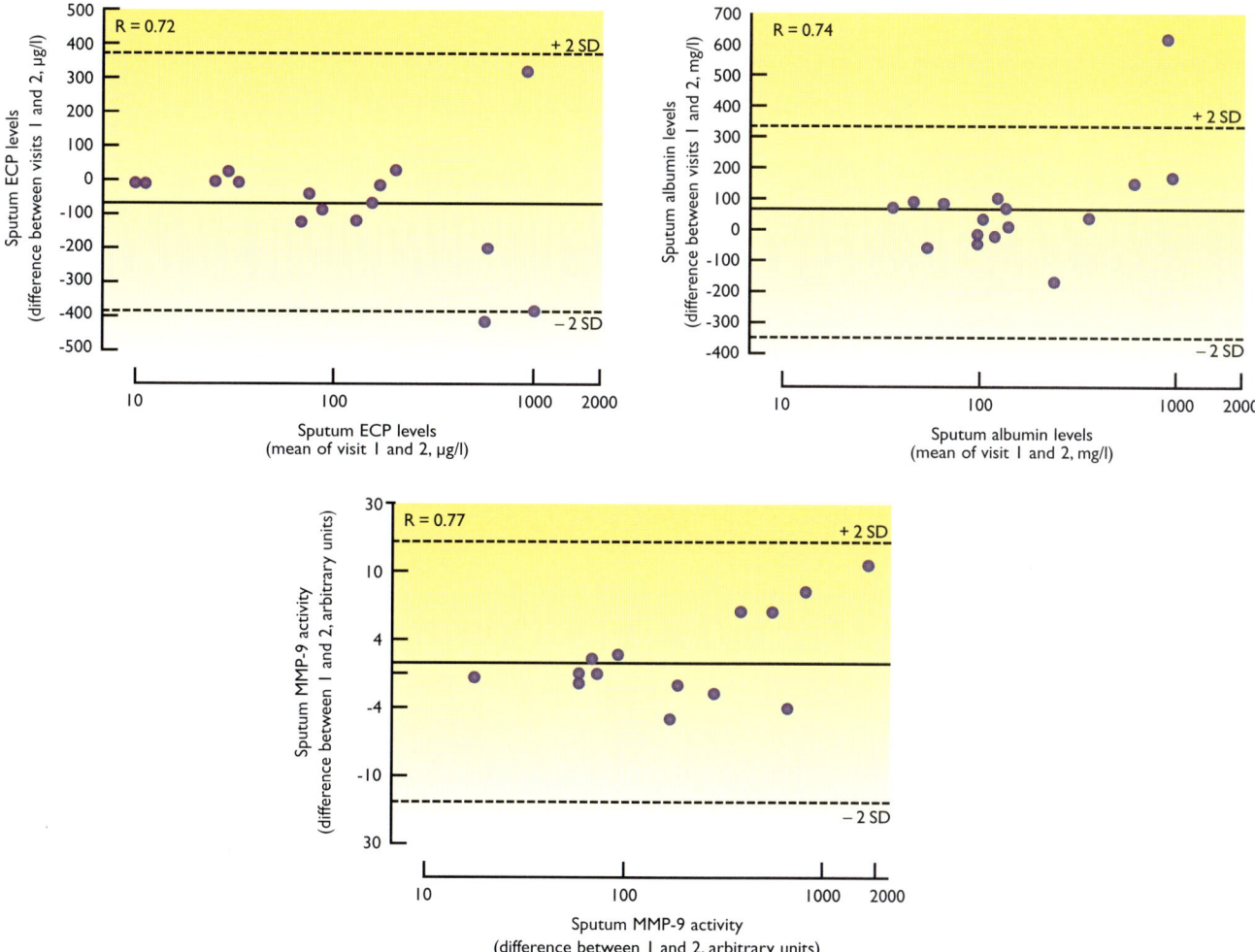

Figure 3.8 Repeatability of eosinophil cationic protein (ECP), albumin, and matrix metalloproteinase (MMP) measurements. This was assessed in 16 subjects who underwent sputum induction twice at weekly intervals with isotonic or hypertonic saline. A Bland–Altman graph is shown where, if the variable is reproducible, it is expected that 95% of the differences between measures are less than 2 standard deviations (SD). For each measured mediator, the differences between the two measurements are plotted against the mean value of the two measurements. There was no difference in the levels of mediators in sputum induced by isotonic saline compared with induction by hypertonic saline. In addition, the repeatability of the measurements was good, as measured by the global intraclass coefficient (R > 0.7). Adapted with permission from Cataldo D, Foidart JM, Louis R, et al. Induced sputum: comparison between isotonic and hypertonic saline solution inhalation in patients with asthma. *Chest* 2001;120:1815–21

of ECP, albumin and pro-matrix metalloproteinase (pro-MMP)-9 (Figure 3.8), but it must not be assumed that this applies to all other, as yet, unmeasured mediators whose release may be modulated by inhalation of hypertonic saline.

FACTORS RELATED TO PREPARATION AND PROCESSING OF SPUTUM

However much precaution is taken, samples obtained during induction will inevitably contain salivary contamination. In order to minimize contamination, some investigators select the mucoid portions of the sample that appear opalescent. Several studies have compared the two ways of collecting sputum and have found that whether sputum is selected or the entire expectorate is used may have an effect on the concentration of mediator, with ECP being found to be higher in selected sputum (Figure 3.9).

Saliva itself has been shown to have low levels of ECP, histamine, tryptase, albumin and fibrinogen

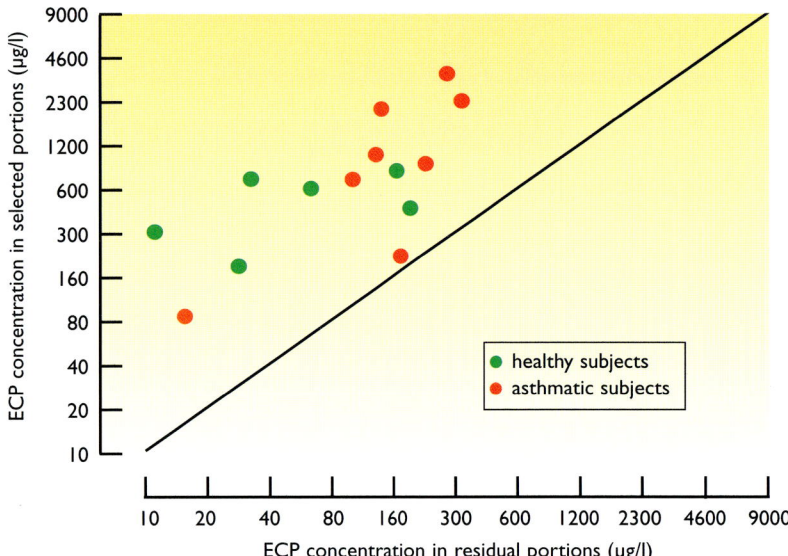

Figure 3.9 Concentrations of eosinophil cationic proteins (ECP) in supernatants from the selected, viscous portion of sputum and the non-viscous portions of the same samples that remained after selection (residual portion). In all subjects, the ECP concentration was greater in the selected portion than in the residual portion ($p < 0.001$, groups comparison). Adapted with permission from Pizzichini E, Pizzichini MM, Efthimiadis A, *et al.* Indices of airway inflammation in induced sputum: reproducibility and validity of cell and fluid-phase measurements. *Am J Respir Crit Care Med* 1996;154:308–17

compared with sputum (Figure 3.10); hence contamination of sputum by saliva probably does not contribute significantly to the levels of these mediators when the whole expectorate is analyzed. The dilutional effect of salivary contamination may be minimized by asking subjects undergoing sputum induction to spit out saliva into a separate container before coughing sputum into another. In this way, the concentration of ECP is increased significantly (Figure 3.11). Similarly, whether sputum is selected or the entire expectorate is processed does not appear to affect the correlation of ECP levels between the two methods, which is good in both cases (Figure 3.12).

Mucolytic agents such as dithiothreitol (DTT) or dithioerythritol (DTE) are widely used as efficient sputum dispersal agents and are popular because their use allows examination of the cellular phase and, in some instances, helps with extraction of soluble mediators. Thus inadequate dispersal of sputum – when using saline – reduces the measured levels of ECP (Figure 3.13).

However, DTT may interfere with the disulphide bonds present in numerous mediators and antibodies, with the potential to interfere with immunoassays. DTT, in a final concentration of 5 mM, has been shown to interfere with the measurement of eosinophil peroxidase (EPO) and myeloperoxidase (MPO) (Figure 3.14), as they

both contain disulphide bonds that may be susceptible to the reducing effect of DTT.

DTT has also been shown to interfere with the standard curve for competitive assays, one example being the assay for cysteinyl leukotrienes (Figure 3.15). Similarly, DTT interferes with the enzyme-linked immunosorbent assay (ELISA) for nerve growth factor (NGF), as shown in Figure 16. In contrast, DTT was shown to have no effect on standards for IL-5 (Figure 3.17), despite IL-5 containing disulphide bonds, indicating that the effect of DTT on a mediator is difficult to predict and should be tested for each individual mediator.

The effect of temperature during processing has not been fully evaluated, but a range from 22 to 37°C does not appear to affect levels of histamine, ECP, eosinophil protein X (EPX), EPO or MPO, (Figures 3.14 and 3.18) and processing NGF at 4°C on ice showed no advantage over processing at room temperature (Figure 3.19).

The addition of protease inhibitors to sputum prior to processing has been shown to significantly increase measured levels of NGF and IL-5 (Figures 3.19 and 3.20). This increase in levels is not due to a spurious effect on the standard curve, as shown in Figure 3.16.

Much attention has been paid to the issue of possible effects of tonicity on mediator levels, not

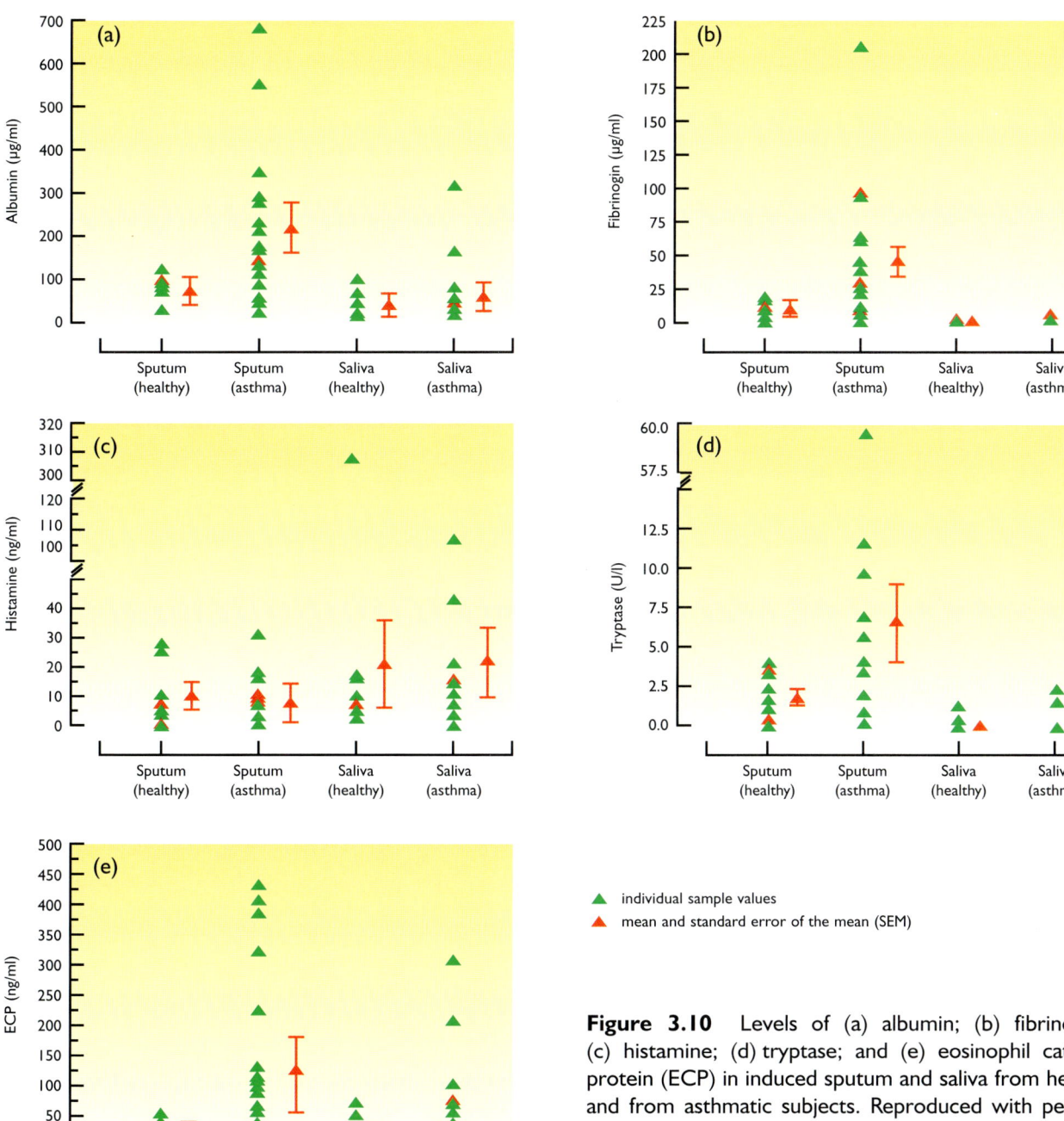

▲ individual sample values

▲ mean and standard error of the mean (SEM)

Figure 3.10 Levels of (a) albumin; (b) fibrinogen; (c) histamine; (d) tryptase; and (e) eosinophil cationic protein (ECP) in induced sputum and saliva from healthy and from asthmatic subjects. Reproduced with permission from Fahy JV, Liu J, Wong H, Boushey HA. Cellular and biochemical analysis of induced sputum from asthmatic and from healthy subjects. *Am Rev Respir Dis* 1993;147:1126–31

least because changes in osmolality can influence cell activity. Increasing the osmolality of processing fluids increases ECP levels in the sputum supernatant, therefore the use of either isotonic or hypertonic fluids to process sputum should be standardized (Figure 3.21).

VALIDATION OF MEASUREMENTS OF SOLUBLE MEDIATORS

Most immunoassays have only been validated for use in blood or cell culture fluid. Sputum is a complex fluid and assays for different mediators should

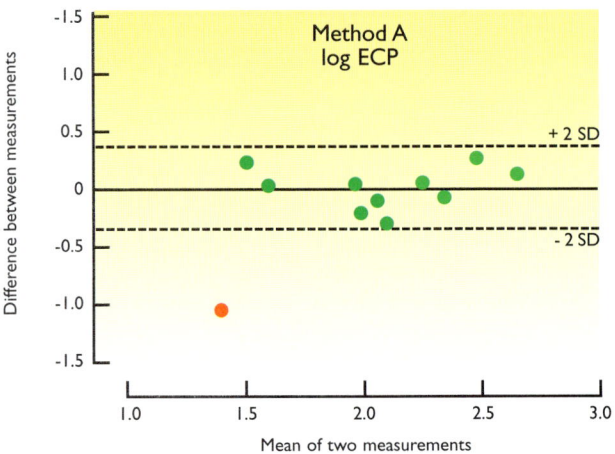
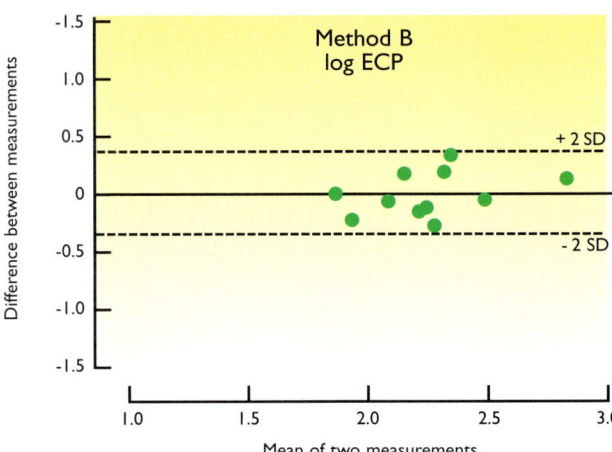

Figure 3.11 Bland–Altman plots for repeated measurements of log transformed eosinophil cationic protein (ECP) levels in induced sputum collected using method A (left panel) and method B (right panel). Method A utilized all secretions, sputum and saliva; method B excluded saliva by asking subjects to spit it out before coughing up sputum. The red circle represent an outlier. Dashed lines represent mean ± 2 standard deviations. Despite the levels of ECP being higher in method B, both methods showed similar reproducibility. Adapted with permission from Gershman NH, Wong HH, Liu MC, *et al*. Comparison of two methods of collecting induced sputum in asthmatic subjects. *Eur Respir J* 1996;9:2448–53

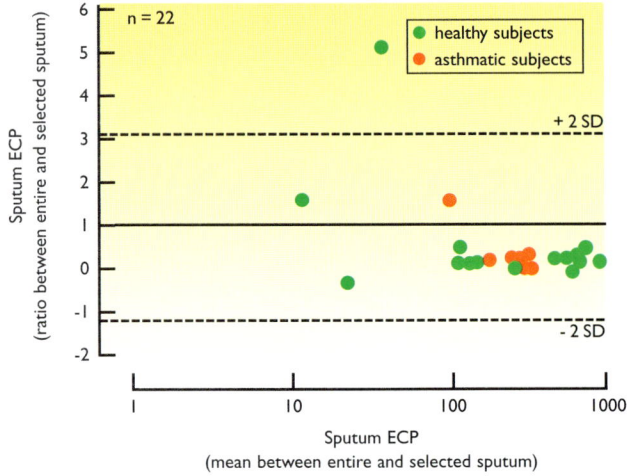

Figure 3.12 Bland–Altman plot of eosinophil cationic protein (ECP) concentration as mean of two values obtained with two different methods (entire and selected sputum) compared with the ratio of the value obtained with the selected and entire method. Both methods have a similar ability to distinguish asthmatics from healthy subjects. SD, standard deviation; n, number of measures plotted. Adapted with permission from Spanevello A, Beghe B, Bianchi A, *et al*. Comparison of two methods of processing induced sputum: selected versus entire sputum. *Am J Respir Crit Care Med* 1998;157:665–8

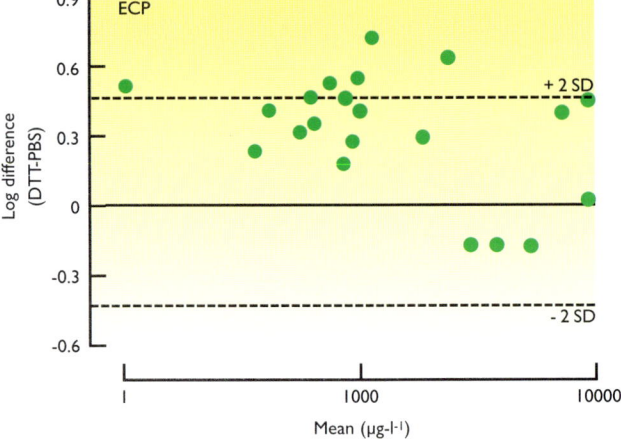

Figure 3.13 Bland–Altman plot of eosinophil cationic proteins (ECP) concentration (log transformed) as mean of two values obtained with sputum dispersed by dithiothreitol (DTT) or by phosphate buffered saline (PBS) compared with the ratio of the value obtained with the two methods. SD, standard deviation. Adapted with permission from Efthimiadis A, Pizzichini MM, Pizzichini E, *et al*. Induced sputum cell and fluid-phase indices of inflammation: comparison of treatment with dtt vs. PBS. *Eur Respir J* 1997;10:1336–40

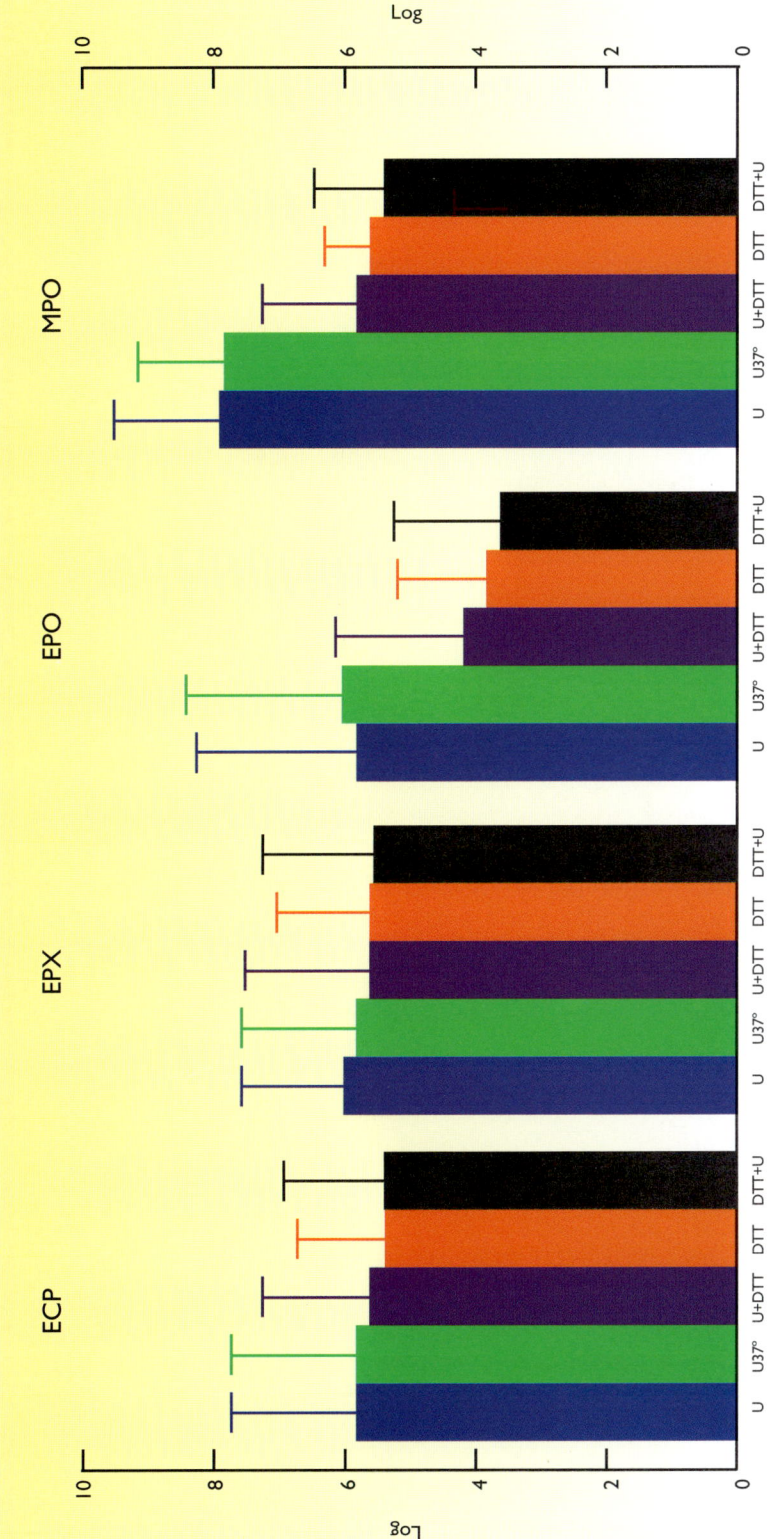

Figure 3.14 Concentrations of eosinophil cationic protein (ECP), eosinophil protein X (EPX), eosinophil peroxidase (EPO), and myeloperoxidase (MPO) in induced sputum that was homogenized by different methods (mean ± standard deviation). It can be seen that the measurement of EPO and MPO was markedly reduced in sputum after dithiothreitol (DTT) was added to the ultrasonically homogenized (U) sample. In addition, homogenizing the sample at 37°C did not affect the levels compared with samples homogenized at room temperature. Reproduced with permission from Grebski E, Peterson C, Medici TC. Effect of physical and chemical methods of homogenization on inflammatory mediators in sputum of asthma patients. *Chest* 2001;119:1521–5

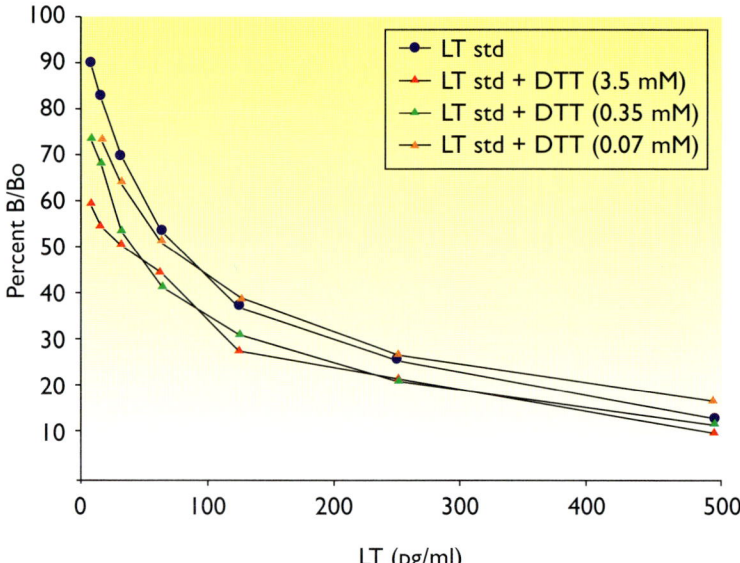

Figure 3.15 The effect of the addition of dithio-threitol (DTT) to the standards in a competitive assay for cysteinyl leukotrienes. Percent B/B0 represents the binding of the competitive inhibitor and is inversely proportional to the leukotriene (LT) concentration. It can be seen that the addition of DTT to the standards results in a shift to the left of the standard curve, which would result in an underestimation of the leukotriene concentration in the assayed samples. The DTT has a dose-dependent effect, with a minimal effect on the curve when the effective concentration of DTT in the standard is 0.07 mM. Reproduced with permission from Kelly MM, Wightman PD, Hargreave FE, *et al*. Repeatability of leukotriene measurements in induced sputum. *Eur Respir J* 2000;16:220S

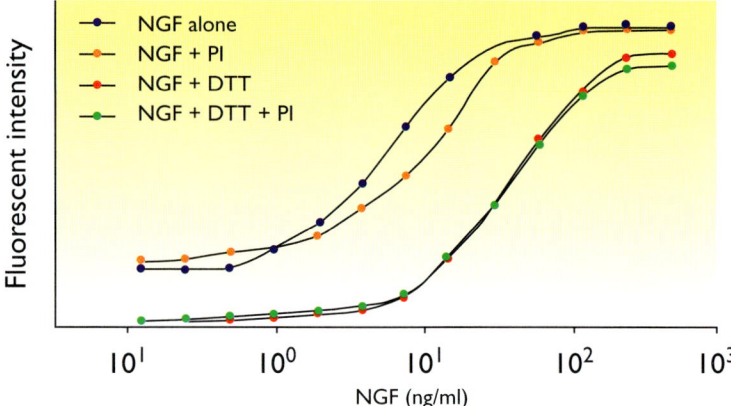

Figure 3.16 The effect of the addition of dithiothreitol (DTT) and/or protease inhibitor (PI) to standards for nerve growth factor (NGF). The curve is shifted to the right, resulting in a false elevation of measured levels of assayed samples. The addition of PI has no effect on the standard curve. Reproduced with permission from Kelly MM, Davis P, Leigh R, *et al*. Measurement of nerve growth factor in sputum. *Am J Respir Crit Care Med* 2001;163: A627

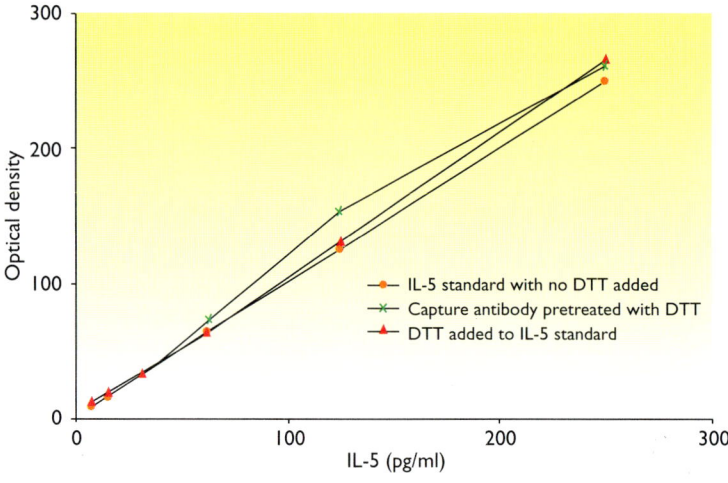

Figure 3.17 The effect of dithiothreitol (DTT) on immunoassay of IL-5 by ELISA. DTT (final concentration 2.5 mM) had no significant effect on the standards even when the capture antibody for the assay was pretreated with DTT. Reproduced with permission from Kelly MM, Leigh R, Hargreave FE, *et al*. Induced sputum: validity of fluid-phase IL-5 measurment. *J Allergy Clin Immunol* 2000;105:1162–8

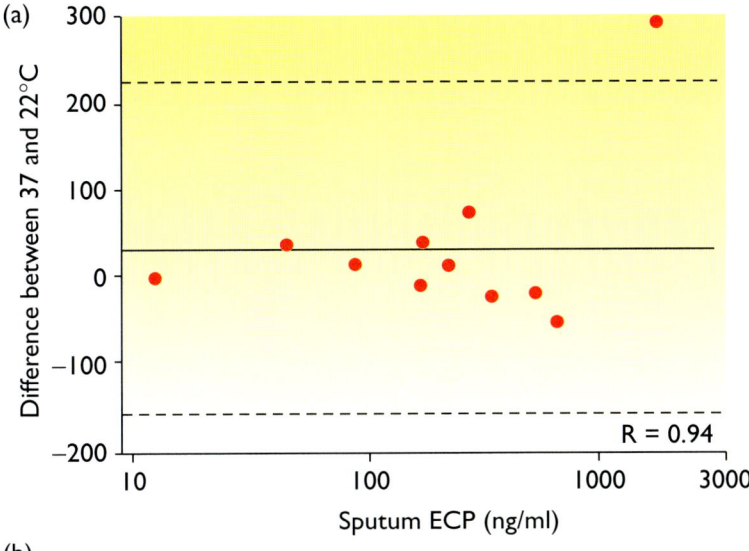

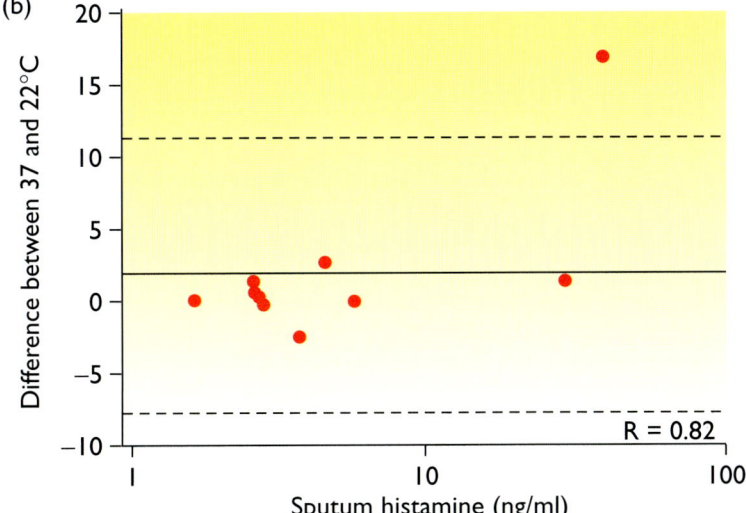

Figure 3.18 Bland–Altman plots for sputum. (a) Eosinophil cationic protein (ECP) and (b) histamine levels when sputum was processed at 37 and 22ÞC in 11 asthmatics. R, intraclass coefficient of correlation. Levels of ECP and histamine were unaffected by the different processing temperatures. Adapted with permission from Louis R, Shute J, Goldring K, et al. The effect of processing on inflammatory markers in induced sputum. *Eur Respir J* 1999; 13:660–7

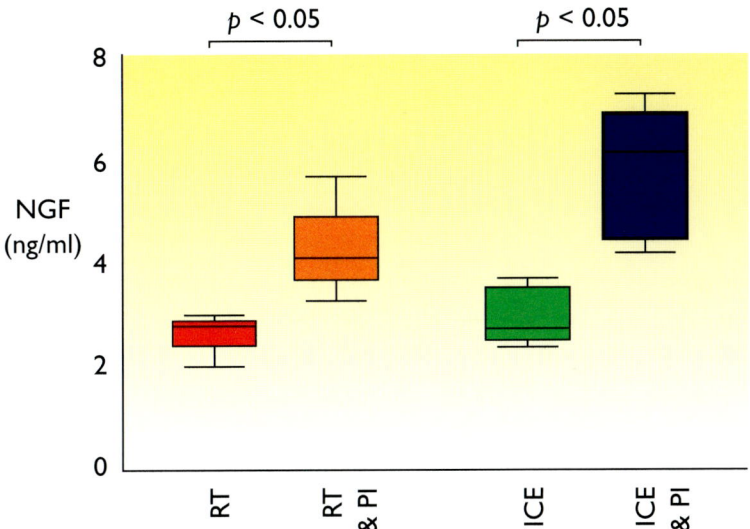

Figure 3.19 Box plot showing nerve growth factor (NGF) levels in samples of sputum from 12 asthmatics and the effect of the addition of protease inhibitors (PI) to sputum prior to processing with dithiothreitol (DTT), as well as the effect of processing at room temperature (RT) or on ice. The addition of PI resulted in significantly increased levels of NGF ($p < 0.05$); there was no advantage of processing the sputum on ice. Reproduced with permission from Kelly MM, Davis P, Leigh R, et al. Measurement of nerve growth factor in sputum. *Am J Respir Crit Care Med* 2001;163:A627

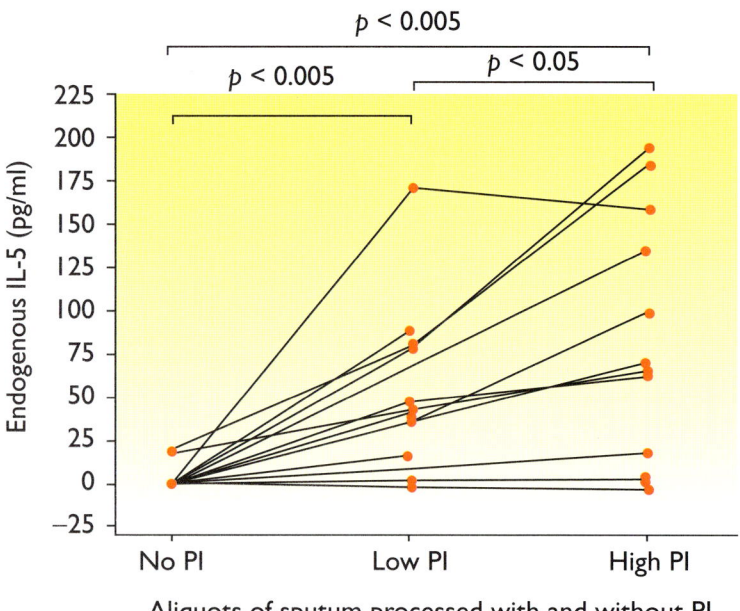

Figure 3.20 Concentration of IL-5 in sputum supernatant processed with no protease inhibitors (No PI), low concentration PI (Low PI) and high concentration PI (High PI). It can be seen that there is a dose-dependent increase in IL-5 measurement with increasing concentration of PI. PI had no effect on the standard curve of the assay (data not shown). Reproduced with permission from Kelly MM, Leigh R, Carruthers S, *et al.* Increased detection of interleukin-5 in sputum by addition of protease inhibitors. *Eur Respir J* 2001;18: 685–91

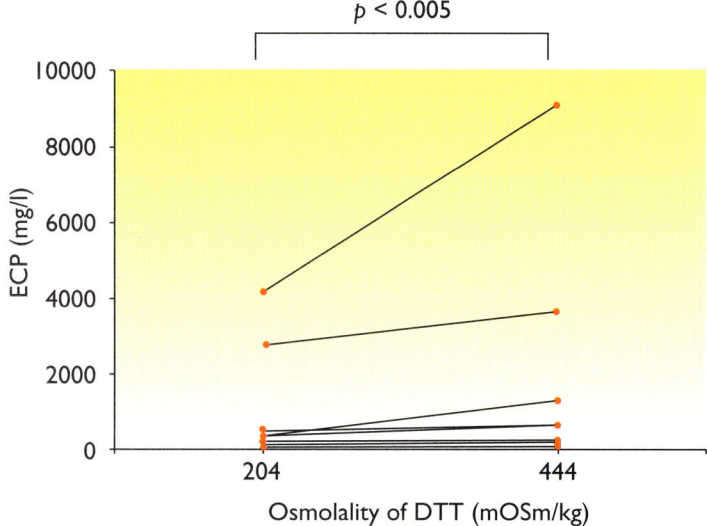

Figure 3.21 Sputum divided into two aliquots and homogenized with dithiothreitol (DTT) diluted with either water (osmolality 204 mOsm/kg) or 0.9% phosphate buffered saline (osmolality 444 mOsm/kg). Eosinophil cationic protein (ECP) measured by radioimmunoassay in sputum supernatant is significantly higher in the aliquot processed with hypertonic DTT. Reproduced with permission from Kelly MM, Davies P, Hargreave FE, Cox G. Osmolality of dispersed sputum: effect of diluent fluids. *Am J Respir Crit Care Med* 2000; 161:A854

be validated individually. A useful method to establish the validity of measurements involves the process of 'spiking' (Figure 3.22). This consists of adding a known amount of the mediator of interest to sputum (the 'spike') before processing, and then measuring the concentration of the mediator in spiked and unspiked aliquots. The spiked mediator should be identical to the standard used in the assay and the percentage recovery of the spike can then be estimated.

Good recovery of the spiked mediator (> 80%) suggests that the assay is sufficiently valid, and if measured endogenous mediator is low or not detected, this can be assumed to be true. If there is poor recovery however, low/absent levels of mediator could represent false negatives. Since various

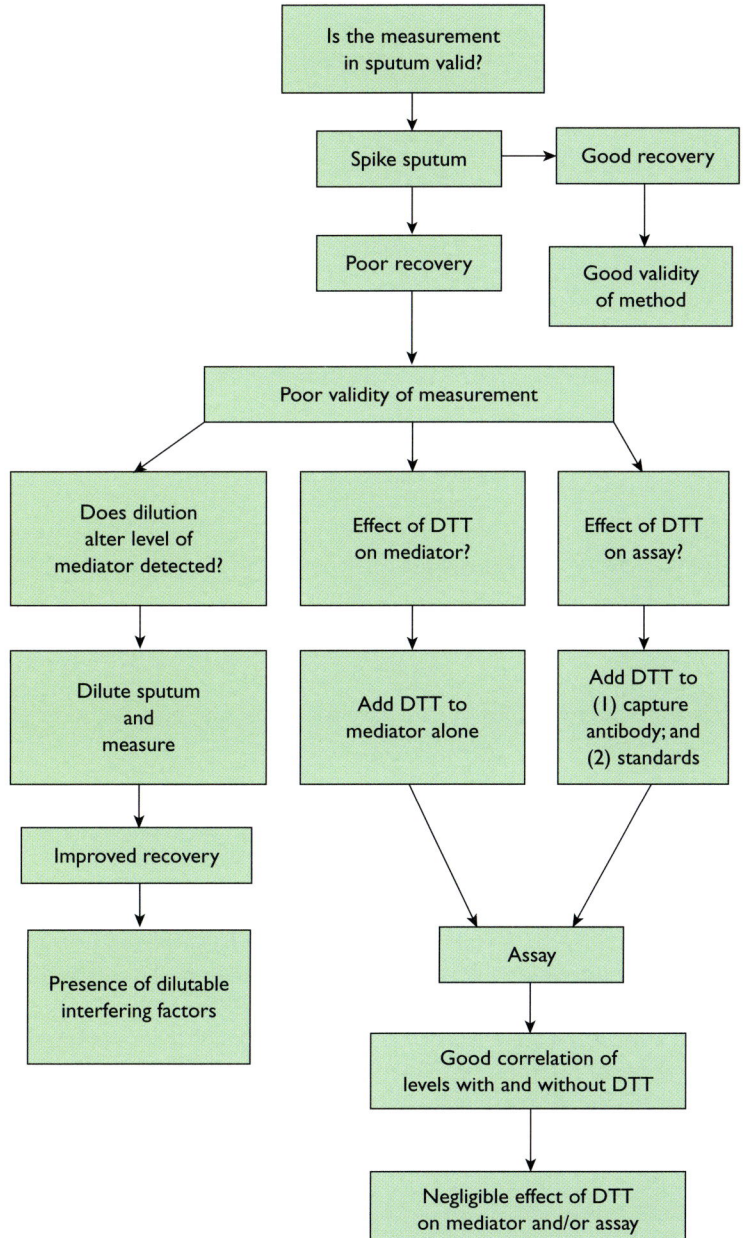

Figure 3.22 Suggested steps to ensure validity of an assay for a sputum mediator. Reproduced with permission from Kelly MM, Leigh R, Hargreave FE, *et al.* Induced sputum: validity of fluid-phase IL-5 measurement. *J Allergy Clin Immunol* 2000;105:1162–8

substances can interfere with the detection of the mediator of interest, including autoantibodies, soluble receptors, and α_2-macroglobulin, dilution studies can be performed to detect interfering substances. The concentration of mediator measured after dilution should not be significantly different (after correcting for dilution), and if it is increased, this suggests the presence of dilutable interfering substances. The supernatant may be diluted with or without the addition of a spike.

REPEATABILITY

The repeatability or reproducibility of measurement of a mediator is the ultimate test of the usefulness

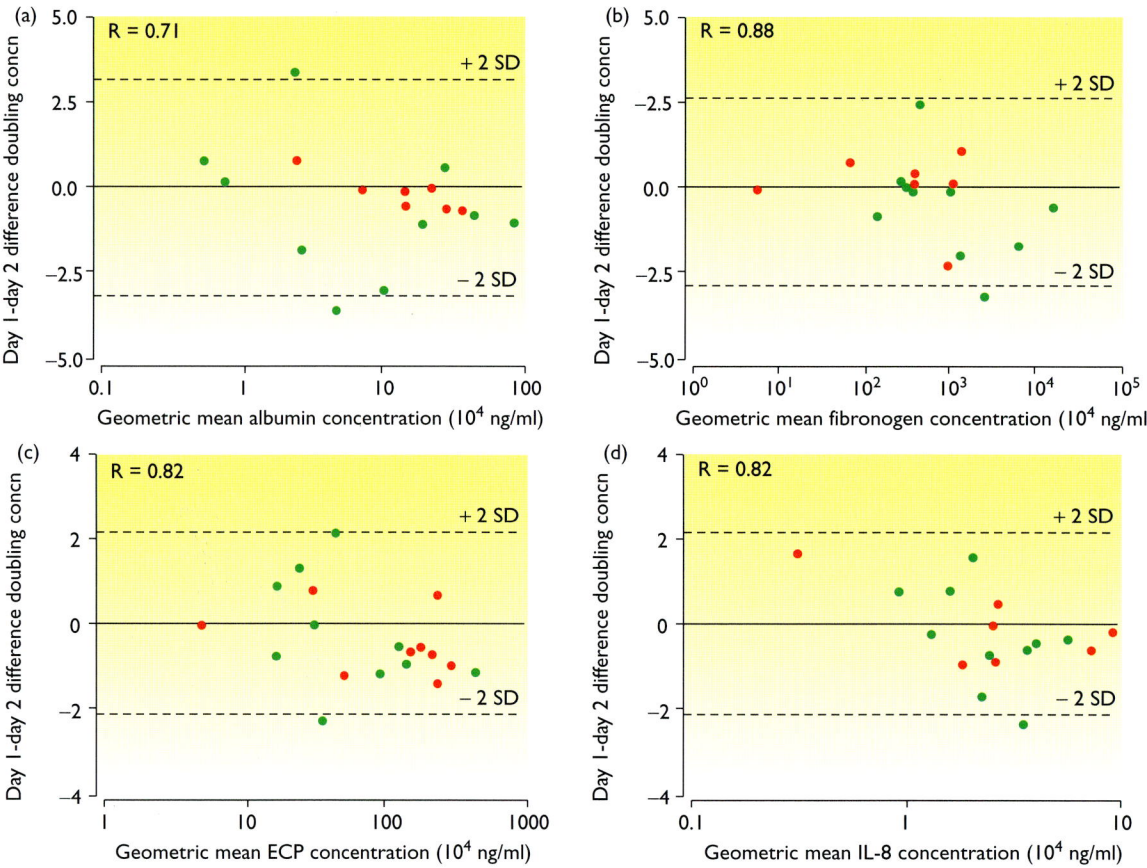

Figure 3.23 Repeatability (Bland–Altman graphs) of (a) albumin; (b) fibrinogen; (c) eosinophil cationic protein (ECP); and (d) interleukin (IL)-8 in induced sputum from patients with mild (green circles) and moderate to severe asthma (red circles). The geometric mean concentration of each subject is plotted against the difference between day 1 and day 2 in doubling concentrations. The solid line represents the line of identity. The limits of agreement (CR) for the whole group are represented by the broken lines (line of identity ± 2 standard deviations (SD) in doubling concentrations). CR, coefficient of repeatability; R, intraclass correlation coefficient. Adapted with permission from In't Veen JCCM, de Gouw HWFM, Smits HH, et al. Repeatability of cellular and soluble markers of inflammation in induced sputum from patients with asthma. Eur Respir J 1996;9:2441–7

of this measurement. Several studies have looked at the repeatability of various mediators, including albumin, fibrinogen, IL-4, -6, -8, tumor-necrosis factor-alpha (TNF-α), ECP, major basic protein (MBP) and eosinophil derived neurotoxin (EDN) (Figures 3.23–3.25). Reproducibility is expressed as proposed by Bland and Altman and it is expected that 95% of the differences between measures are < 2 standard deviations (SD). Generally, reproducibility of mediator levels in induced sputum is good, which allows their use as indicators of responsiveness to treatment – very

useful for clinical trials – and also helps to define inflammatory responses to such triggers of asthma as infections and allergen exposure.

In conclusion, precautions need to be taken when measuring individual mediators, since sputum is a unique biological fluid and cannot be considered equivalent to cell culture supernatant or serum. When these facts are taken into consideration, the measurement of soluble contents in sputum has the potential to be an extremely useful and convenient research tool.

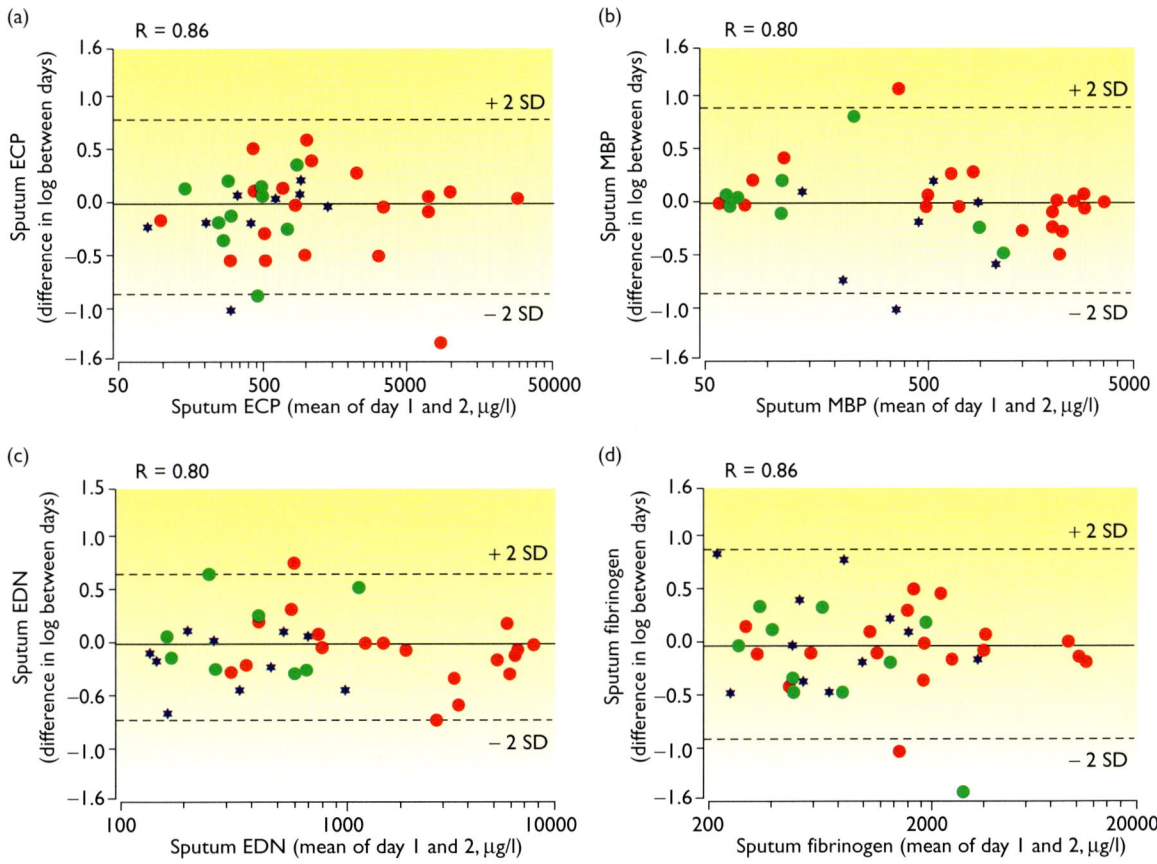

Figure 3.24 Repeatability of (a) eosinophil cationic protein (ECP); (b) major basic protein (MBP); (c) eosinophil derived neurotoxin (EDN); and (d) fibrinogen levels with Bland–Altman graphs. The differences of log values of days 1 and 2 are plotted on the vertical axis against the mean value of the two measurements (days). R is the global intraclass correlation coefficient and the area between the dashed lines is ± 2 standard deviations (SD) of the mean of the two differences. Green circles, healthy subjects; red circles, asthmatics; stars, smokers with non-obstructive bronchitis. Adapted with permission from Pizzichini E, Pizzichini MM, Efthimiadis A, *et al*. Indices of airway inflammation in induced sputum: reproducibility and validity of cell and fluid-phase measurements. *Am J Respir Crit Care Med* 1996;154:308–17

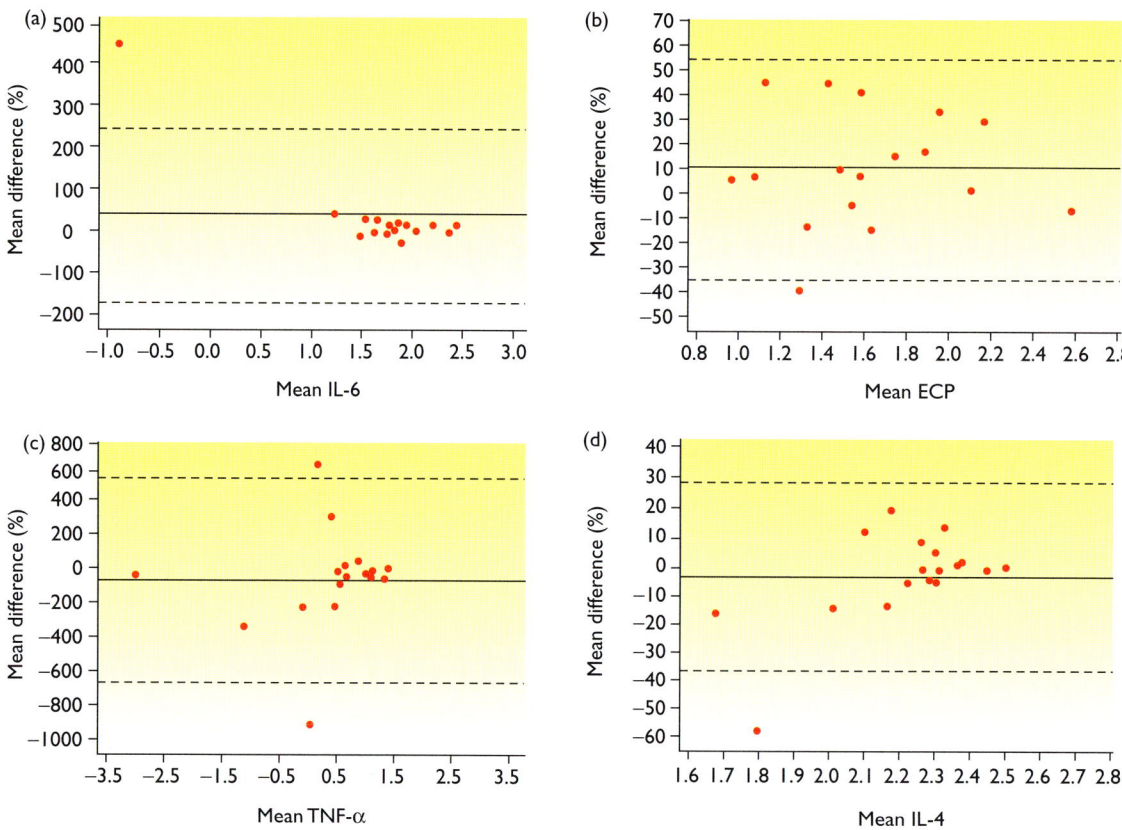

Figure 3.25 Reproducibility of (a) interleukin (IL)-6 (pg/ml); (b) tumor necrosis factor (TNF)-α (pg/ml); (c) IL-4 (pg/ml); and (d) eosinophil cationic protein (ECP) (ng/ml) in induced sputum (percent). Mean represented by solid lines, ± 1.96 standard deviations (dotted lines). Reproducibility of measurements is expressed as proposed by Bland and Altman. Adapted with permission from Purokivi M, Randell J, Hirvonen MR, Tukiainen H. Reproducibility of measurements of exhaled NO, and cell count and cytokine concentrations in induced sputum. *Eur Respir J* 2000;16:242–6

CHAPTER 4

Induced sputum studies in asthma

Renaud A. Louis and Pascal Chanez

Although the technique of sputum induction was first developed to detect infectious agents in the airways of immunocompromised patients, in the past ten years it has been shown to be a very useful tool in the study of airway diseases such as asthma and chronic obstructive pulmonary disease (COPD). Consistent with this, there has been a tremendous expansion in its application in the study of basic mechanisms of airway inflammation and the monitoring of airway inflammation. The first study conducted by Pin and colleagues showed raised numbers of eosinophils and metachromatic cells in the sputum cell fraction from asthmatics compared with healthy subjects (Figure 4.1). Soon afterwards, Fahy and colleagues showed that the fluid phase of the sputum samples contained increased levels of eosinophil cationic protein (ECP) and albumin (Figure 3.10 in Chapter 3) compared with healthy subjects – soluble biochemical markers that can be measured by commercial immunoassays.

Numerous studies have since validated the measurements of both the cellular and the soluble components of relevance to asthma. These have

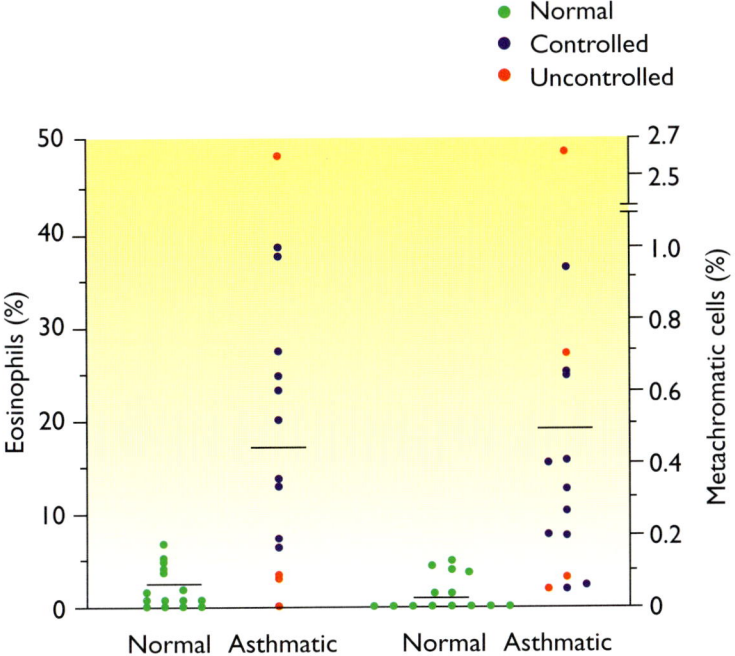

Figure 4.1 Sputum eosinophil (left) and metachromatic (right) cell counts in control (normal) and asthmatic subjects. Among asthmatics the blue circles represent the controlled asthmatics while the red circles are those with controlled asthma. The solid bars represent medians. Reproduced with permission from the BMJ Publishing Group from Pin I, Gibson PG, Dolovich J, *et al*. Use of induced sputum cell counts to investigate airway inflammation in asthma. *Thorax* 1992;47: 25–9

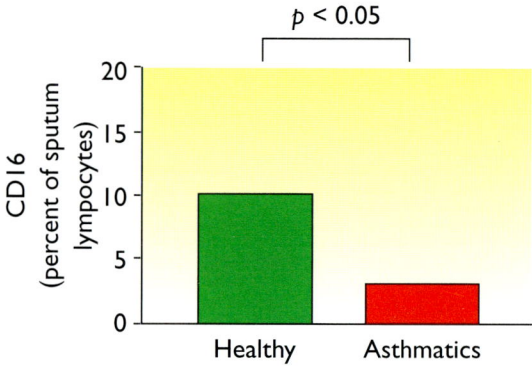

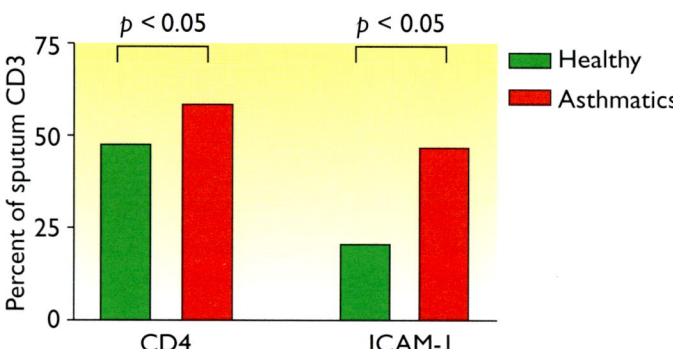

Figure 4.2 Percentages of natural killer cells that are CD16-positive cells within the sputum lymphocyte gate (upper panel) and CD4 cells as well as those expressing the adhesion molecule ICAM-1 shown as a percentage of total CD3+ T cells. Reproduced with permission from Louis R, Shute J, Djukanovic R, et al. Cell infiltration, ICAM-1 expression, and eosinophil chemotactic activity in asthmatic sputum. Am J Respir Crit Care Med 1997;155:466–72

helped understand the underlying inflammatory processes that are central to the pathophysiologic and clinical features of asthma both in stable disease and during exacerbation. Several large cross-sectional studies and intervention studies have been conducted; these have cast light on the relationship between airway inflammation and the clinical expression of the disease, including bronchial hyperresponsiveness. More recently, there has been significant interest in the use of induced sputum for monitoring asthma and optimizing disease control.

AIRWAY INFLAMMATION IN CHRONIC STABLE ASTHMA

Together with mast cells and eosinophils, airway infiltration with T-helper type 2 (Th2) CD4+ T lymphocytes is a central feature of asthma pathophysiology. Although T lymphocytes are less numerous in sputum than in bronchial biopsies or bronchoalveolar lavage (BAL), and usually represent less than 3% of sputum cells, these cells have been successfully phenotyped using flow cytometry. Asthmatics have a slightly raised number of CD4+ T cells expressing surface activation markers such as intercellular adhesion molecule-1 (ICAM-1) (Figure 4.2). Studies have also shown a reduced proportion of lymphocytes expressing the marker CD16 identifying these as natural killer (NK) cells. Raised B lymphocyte counts have been shown to correlate with sputum eosinophil counts (Figure 4.3). In keeping with this, increased amounts of secretory immunoglobulin A (IgA) have been detected in the sputum of asthmatics, which forms complexes with interleukin (IL)-8. These correlate with the number of sputum eosinophils and their extent of activation (Figure 4.4). Further interest in IgA in asthma has arisen from a study showing that asthmatics had higher levels of

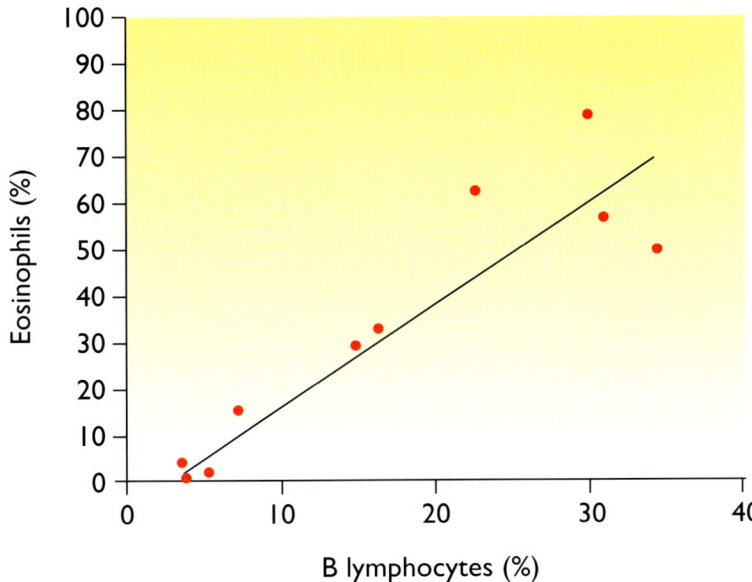

Figure 4.3 Correlation between sputum CD19-positive lymphocytes (B cells) and sputum eosinophil counts in asthmatics. B lymphocytes are expressed as percentages of total sputum lymphocytes; sputum eosinophils are expressed as percentages of total sputum cell counts. Reproduced with permission from Kidney JC, Wong AG, Hargreave FE, *et al*. Elevated B cells in sputum of asthmatics. Close correlation with eosinophils. *Am J Respir Crit Care Med* 1996;153: 540–4

sputum IgA that were not only directed towards mite allergen but also towards Streptococcus pneumoniae antigen (Figure 4.5). Interestingly, the IgA levels correlated with the extent of local eosinophil activation, as reflected in sputum ECP levels, lending support to the hypothesis that secretory IgA may act as a potent activator for airway eosinophils.

Sputum analysis has also proven to be valid for measuring several inflammatory mediators, among which cysteinyl-leukotrienes deserve particular attention because of their confirmed role in asthma. Pavord and colleagues have found that levels of cysteinyl-leukotrienes are related to asthma severity, as judged by the treatment necessary to control the disease, and further increased in acute asthma (Figure 4.6).

The measurement of cytokines in supernatants from sputum is fraught with problems because the mucolytic used for processing the sputum may interfere with the immunoassays. Some authors successfully overcome this drawback by diluting samples with PBS instead of using a mucolytic. Compared with healthy subjects or asymptomatic asthmatics, symptomatic asthmatics were found to have raised levels of several cytokines, including IL-1b, IL-5, IL-6, tumor necrosis factor (TNF)-α RANTES and IL-8 (Figure 4.7), with no

disease-related change in levels of interferon (IFN)-γ and IL-2. Interestingly, even asymptomatic asthmatics exhibit increased levels of IL-1β, IL-6, IL-5, RANTES and IL-8 compared with healthy subjects. The raised levels of this set of cytokines may contribute to lymphocyte proliferation and granulocyte recruitment and activation, all features of airway inflammation in asthma. In contrast, sputum levels of the anti-inflammatory and immunoregulatory cytokine IL-10 were reduced in asthmatic compared with healthy subjects (Figure 4.8). Recent studies have suggested that elevated secretion of lipoxin A_4, a lipid-derived mediator endowed with anti-inflammatory properties, might protect mild asthmatics against the most severe form of the disease by inhibiting IL-8 production and thereby reducing the extent of the neutrophilic inflammation associated with severe asthma (Figure 4.9).

There has been considerable interest in the mechanisms involved in the remodelling of asthmatic airways. Matrix metalloproteinases (MMPs) have been viewed as being of paramount importance in degrading the extracellular matrix within bronchial walls. Some MMPs are readily detectable in sputum by zymography. Using this technique, moderate to severe asthmatics have been

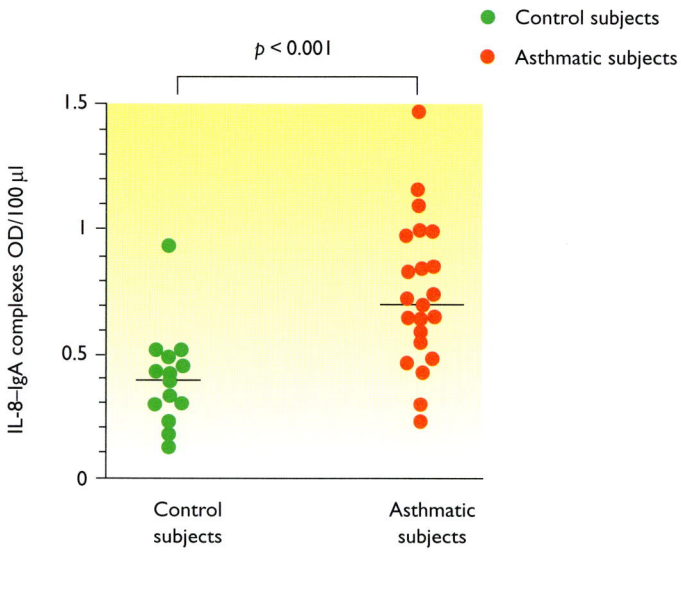

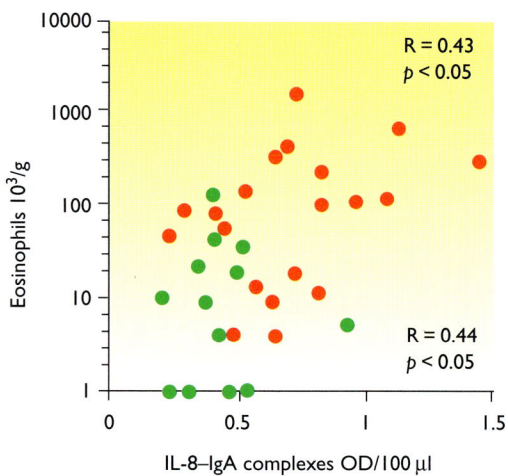

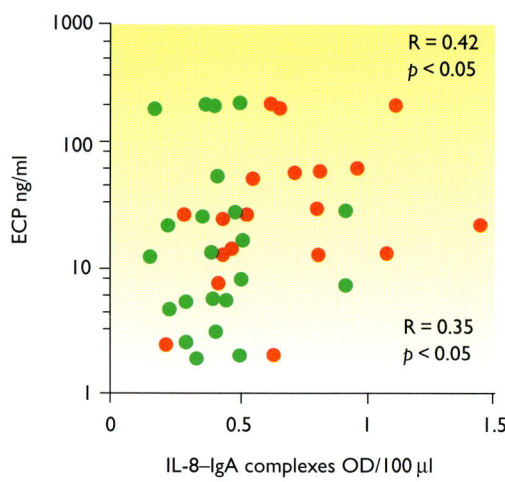

Figure 4.4 The association between IL-8–IgA complexes and eosinophil counts and activation. A comparison of sputum levels of IL-8–IgA in healthy and asthmatic patients is shown in the upper panel. The relationship between sputum levels of IL-8–IgA complexes and eosinophil counts in healthy and asthmatic subjects is shown in the middle panel and ECP levels in the lower panel. Horizontal bars denote median values; R, coefficient of correlation for pooled data; OD, optical density. Reproduced with permission from Louis R, Shute J, Djukanovic R, *et al*. Cell infiltration, ICAM-1 expression, and eosinophil chemotactic activity in asthmatic sputum. *Am J Respir Crit Care Med* 1997;155:466–72

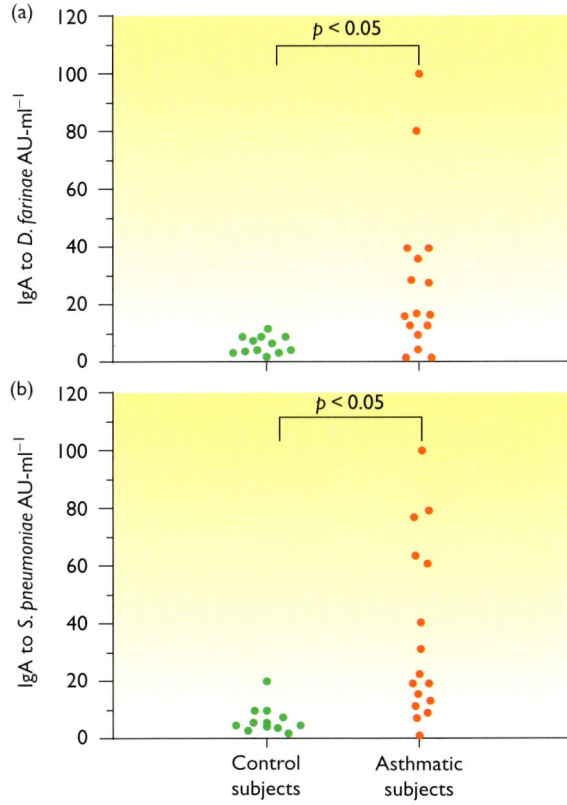

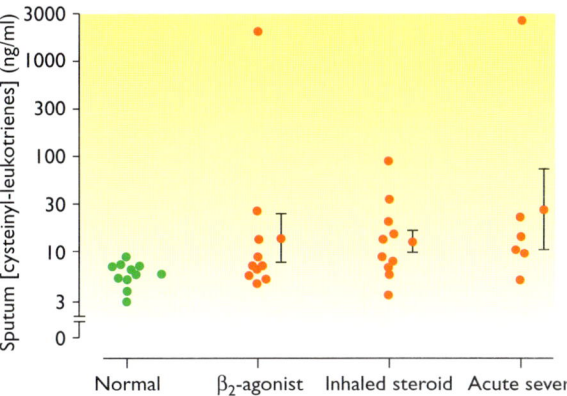

Figure 4.6 Sputum cysteinyl-leukotriene levels in healthy subjects, mild asthmatics treated with β_2-agonists only, moderate asthmatics treated with high doses of inhaled steroids and asthmatics in acute severe exacerbation. Reproduced with permission from Pavord ID, Ward R, Dworski R, *et al.* Induced sputum eicosanoid concentrations in asthma. *Am J Respir Crit Care Med* 1999;160:1905–9

Figure 4.5 Sputum IgA specific for *Dermatophagoides farinae* (a) and *Streptococcus pneumoniae* (b) in control subjects and asthmatics. Reproduced with permission from Nahm DH, Kim HY, Park HS. Elevation of specific immunoglobulin A antibodies to both allergen and bacterial antigen in induced sputum from asthmatics. *Eur Respir J* 1998;12:540–5

found to exhibit greater gelatinolytic activity in their sputum than healthy subjects (Figure 4.10). Asthmatics have also been shown to have more MMP-2, an MMP able to stimulate smooth muscle proliferation.

AIRWAY INFLAMMATION DURING EXACERBATION

As a relatively non-invasive tool, sputum induction has allowed the collection of airway samples during asthma exacerbation, which has not been possible using bronchoscopy. Sputum from asthmatics admitted to hospital for an asthma attack contains high numbers of neutrophils, pointing to the involvement of these cells in the pathogenesis of acute asthma attack (Figure 4.11). Severe exacerbations can also be characterized by high eosinophil counts that fall after treatment with systemic corticosteroids, and the changes are paralleled by improvement in lung function (Figure 4.12). Exacerbations that occur when the dose of inhaled steroids in otherwise well-controlled asthmatics is tapered have also been shown to be associated with a sharp rise in sputum eosinophilia. Furthermore, the exacerbations occurred in those asthmatics displaying a high sputum eosinophil count at the start of reduction of the dose of inhaled steroids (Figure 4.13). In contrast, exacerbations occurring in difficult-to-control asthma were found to be unrelated to a rise in sputum eosinophilia (Figure 4.14).

MECHANISMS UNDERLYING PERSISTENT SPUTUM EOSINOPHILIA

The fluid phase of sputum in asthmatics contains biological activities that may contribute to the raising and maintenance of high eosinophil counts. Theoretically, raised eosinophilia may result either

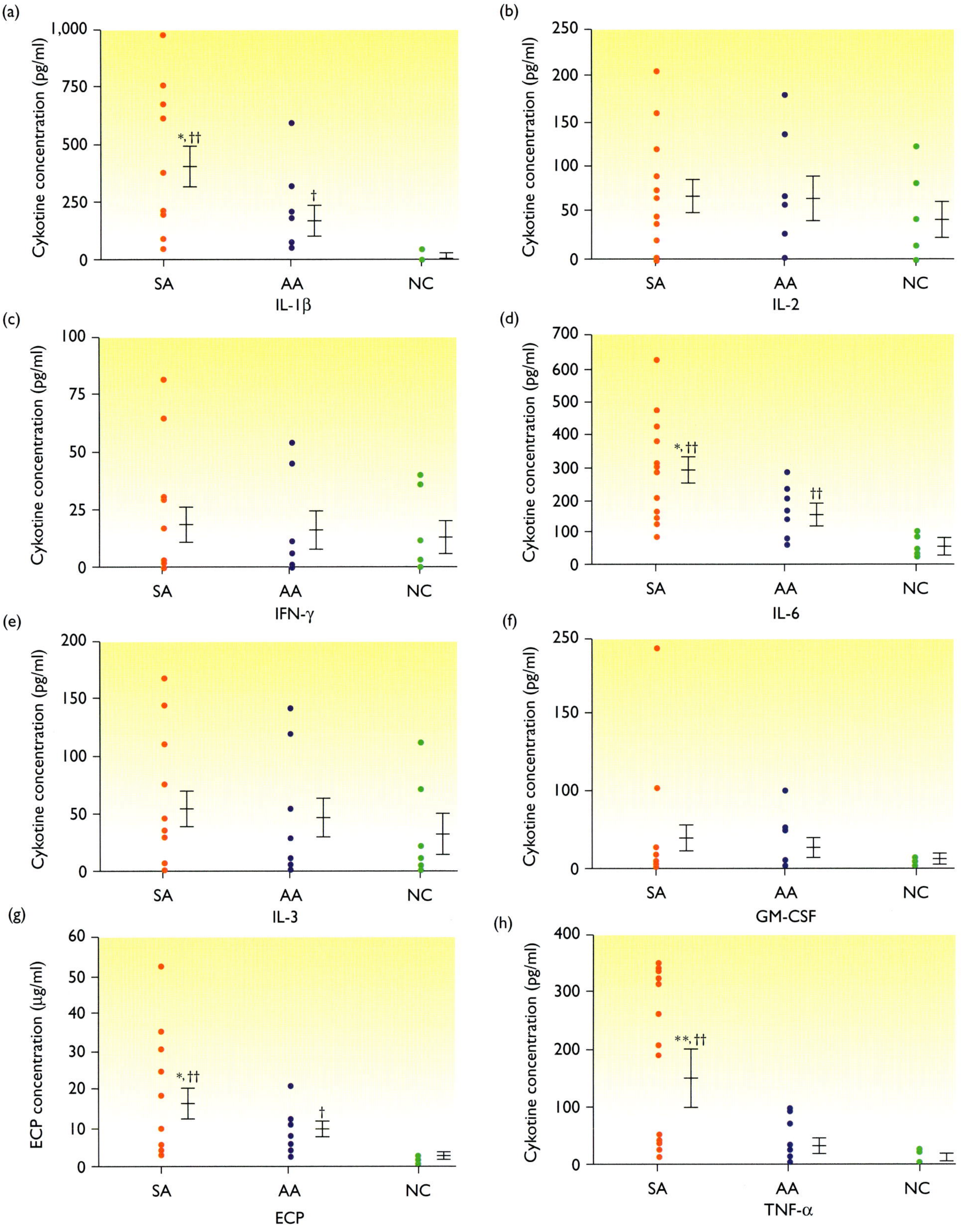

(a)
Cykotine concentration (pg/ml)
1,000
750
500
250
0
*,††
†
SA
AA
NC
IL-1β
(b)
Cykotine concentration (pg/ml)
250
200
150
100
50
0
SA
AA
NC
IL-2
(c)
Cykotine concentration (pg/ml)
100
75
50
25
0
SA
AA
NC
IFN-γ
(d)
Cykotine concentration (pg/ml)
700
600
500
400
300
200
100
0
*,††
††
SA
AA
NC
IL-6
(e)
Cykotine concentration (pg/ml)
200
150
100
50
0
SA
AA
NC
IL-3
(f)
Cykotine concentration (pg/ml)
250
150
100
0
SA
AA
NC
GM-CSF
(g)
ECP concentration (μg/ml)
60
50
40
30
20
10
0
*,††
†
SA
AA
NC
ECP
(h)
Cykotine concentration (pg/ml)
400
300
200
100
0
**,††
SA
AA
NC
TNF-α

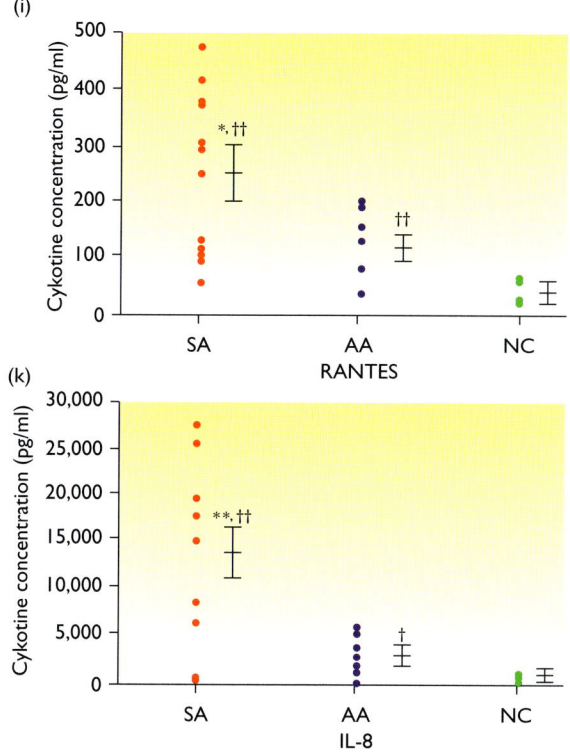

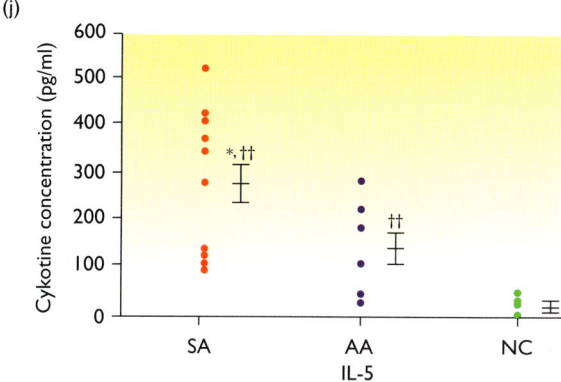

Figure 4.7 Sputum levels of cytokines in symptomatic asthmatics (SA), asymptomatic asthmatics (AA) and normal controls (NC). *$p < 0.05$ and **$p < 0.01$ versus AA; †$p < 0.05$ and ††$p < 0.01$ versus normal controls. Reproduced with permission from S. Karger AG, Basel, from Konno S, Gonokami Y, Kurokawa M, Adachi M, et al. Cytokine concentrations in sputum of asthmatic patients. Int Arch Allergy Immunol 1996;109:73–8

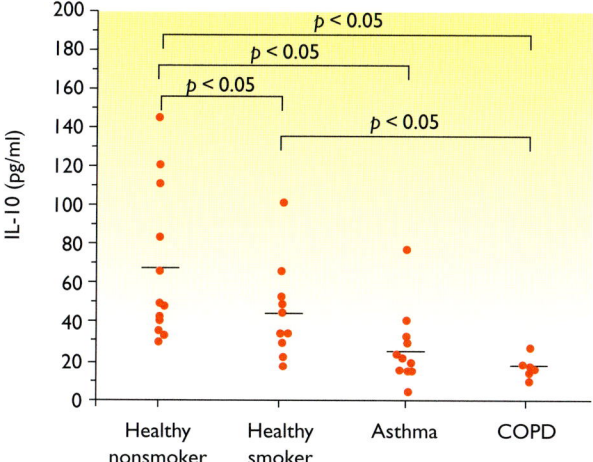

Figure 4.8 Sputum levels of IL-10 in healthy non-smokers, healthy smokers, asthmatics, and patients with COPD. Horizontal bars represent mean values. Reproduced with permission from Takanashi S, Hasegawa Y, Okamura K, et al. Interleukin-10 level in sputum is reduced in bronchial asthma, COPD and in smokers. Eur Respir J 1999;14:309–14

from prolonged survival or increased influx of newly recruited eosinophils. The technique of sputum induction has proved useful in elucidating the mechanisms that may result in bronchial eosinophilia. Sputum from asthmatics collected during an exacerbation possesses anti-apoptotic activity that is partly related to the presence of granulocyte–macrophage colony stimulating factor (GM-CSF) and IL-5 (Figure 4.15). Cells in sputum

from asthmatics have also been shown to express greater levels of cellular inhibitor apoptosis protein-2 (c-IAP2) that strongly correlate with sputum eosinophil counts (Figure 4.16). c-IAP2 is an anti-apoptotic protein whose expression is enhanced following activation of eosinophils via the CD40 accessory molecule. Blood eosinophils do not normally express c-IAP2 regardless of whether they are from healthy control subjects or

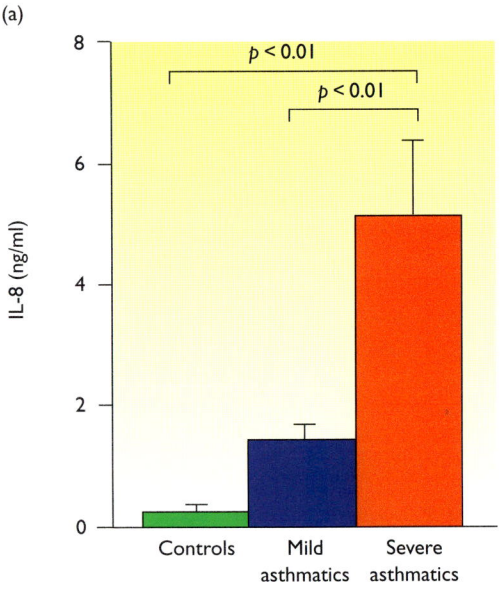
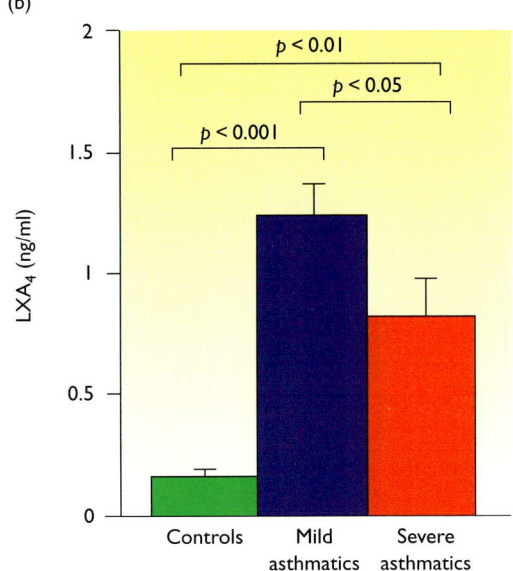

Figure 4.9 Sputum levels of interleukin (IL)-8 and LXA$_4$ in controls and mild and severe asthmatics. While IL-8 levels increase with disease severity, LXA$_4$ levels are lower in severe than in mild asthma, but higher than in healthy subjects. Reproduced with permission from Bonnans C, Vachier I, Chanez P, *et al.* Lipoxins are potential endogenous antiinflammatory mediators in asthma. *Am J Respir Crit Care Med* 2002;165:1531–5

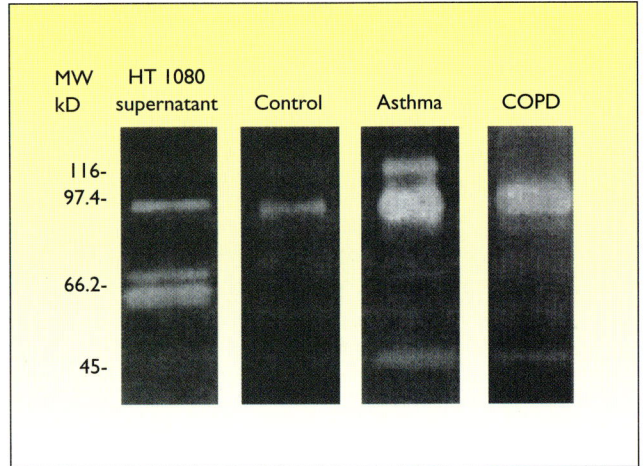
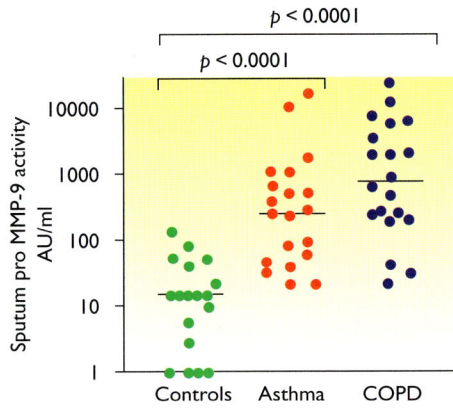

Figure 4.10 Zymographic analysis of sputum supernatants. On the left, the first lane corresponds to supernatant from cultured HT 1080 cells used as reference. HT 1080 cells are known to spontaneously produce high amounts of matrix metalloproteinase (MMP)-9 (gelatinase A) and MMP-2 (gelatinase B). Although the major sputum gelatinolytic activity is displayed at 92 kDa, corresponding to the pro-MMP-9, clear lysis bands were sometimes present at 85 kDa (activated form of MMP-9), 72 kDa (MMP-2), 120 kDa and 45 kDa. On the right is the quantitative analysis of the 92 kDa lysis band performed by densitometric scanning of the gels. Reproduced with permission from S. Karger AG, Basel, from Cataldo D, Munaut C, Louis R, *et al.* MMP-2- and MMP-9-linked gelatinolytic activity in the sputum from patients with asthma and chronic obstructive pulmonary disease. *Int Arch Allergy Immunol* 2000;123:259–67

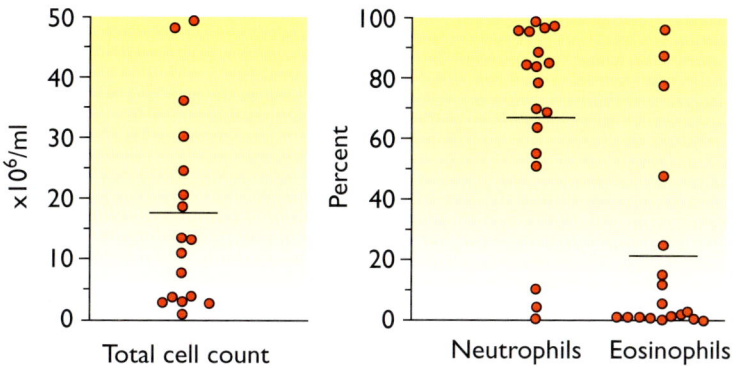

Figure 4.11 Total cell counts and percentages of neutrophils and eosinophils in sputum from 18 subjects with asthma in acute exacerbation admitted to emergency room. The bars represent mean of data. Reproduced with permission from Fahy JV, Kim KW, Liu J, Boushey HA. Prominent neutrophilic inflammation in sputum from subjects with asthma exacerbation. *J Allergy Clin Immunol* 1995;95: 843–52

asthmatic donors. In contrast, c-IAP2 expression is increased in sputum cells from asthmatics. Besides mechanisms that delay eosinophil death, the asthmatic process may involve processes that promote recruitment of blood eosinophils. Consistent with this, significant eosinophil chemotactic activity could be demonstrated in the fluid phase of sputum from asthmatics, a phenomenon not observed in healthy subjects (Figrue 4.17). Among the chemotactic factors potentially involved in asthma is eotaxin, the levels of which have been found to be higher in both stable and unstable asthmatics (Figure 4.18) than non-asthmatics. The source of eosinophil-active chemokines is unknown; mast cells might play a role as suggested by the association between mast-cell activation and the extent of sputum eosinophilia in asthma (Figure 4.19).

RELATIONSHIP BETWEEN SPUTUM EOSINOPHILIA AND INDICES OF ASTHMA SEVERITY

By conducting a large cross-sectional study in 74 asthmatics classified according the GINA criteria for symptoms and airways calibre, Louis and colleagues have shown that sputum eosinophil counts and levels of ECP and albumin are related to the severity of asthma as judged by NHLBI/ World Health Organization (WHO) criteria (Figure 4.20). Using a similar approach to classify asthmatics, another study has related the disease severity to neutrophil counts (Figure 4.21). These

studies have suggested that the persistence of both eosinophilic and neutrophilic inflammation in moderate to severe asthma, despite treatment with high doses of inhaled and sometimes oral steroids, reflects a degree of corticosteroid resistance.

Sputum induction has provided an opportunity to examine the controversial relationship between bronchial hyperresponsiveness and airway eosinophilia in asthma. In a series of 118 steroid-naive asthmatics with normal baseline lung function ($FEV_1 > 70\%$ predicted) prospectively recruited from an outpatient clinic, sputum eosinophil counts were inversely and significantly associated with methacholine bronchial hyperresponsiveness, explaining up to 16% of the variation in the provocative concentration of methacholine causing a 20% fall in FEV_1 ($PC_{20}M$) after multiple regression analysis (Figure 4.22). Over the past few years, there has been a growing interest in bronchial hyperresponsiveness to indirect stimuli, i.e. agents that cause bronchospasm indirectly through interaction with resident inflammatory cells that, in turn, release direct bronchoconstrictor mediators. In a large series of steroid-naive asthmatics, bronchial responsiveness to adenosine – a mediator thought to interact with mast cells – was stronger than to methacholine (Figure 4.23).

Although airway eosinophilic inflammation is broadly related to asthma severity, the precise role of eosinophils in the clinical expression of asthma remains uncertain. Controversy has been generated following a clinical trial in which

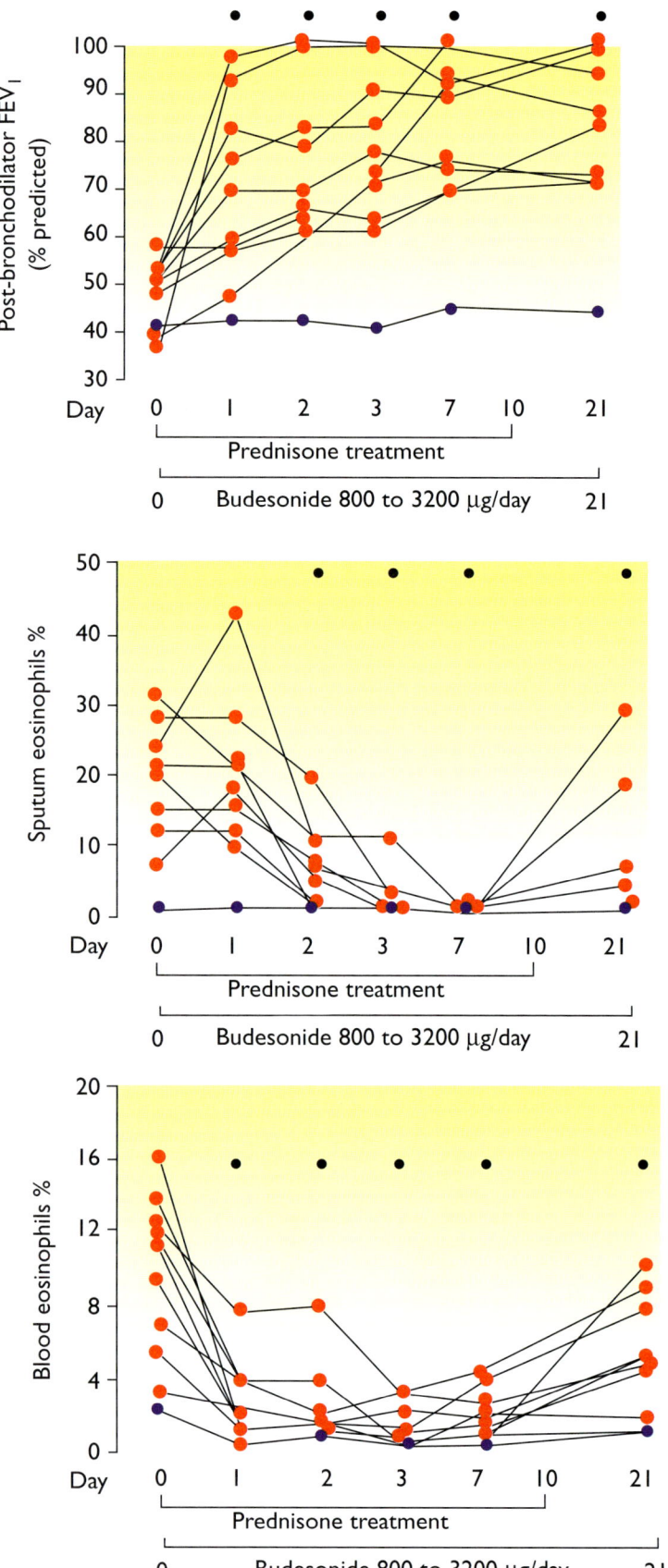

Figure 4.12 Post-bronchodilator forced expiratory volume (FEV₁) values and sputum and peripheral blood eosinophil counts before and after treatment in asthmatics presenting during acute exacerbation. Treatment consisted of oral prednisone 30 mg daily for 5 days followed by reduction to zero by day 10, and inhaled budesonide at doses ranging from 800 to 3200 µg/24 h until day 21. It is worth noting that the only patient whose FEV₁ did not improve was the one with no sputum eosinophils at day 0. Reproduced with permission from Pizzichini MM, Pizzichini E, Hargreave FE, *et al.* Sputum in severe exacerbations of asthma: kinetics of inflammatory indices after prednisone treatment. *Am J Respir Crit Care Med* 1997;155:1501–8

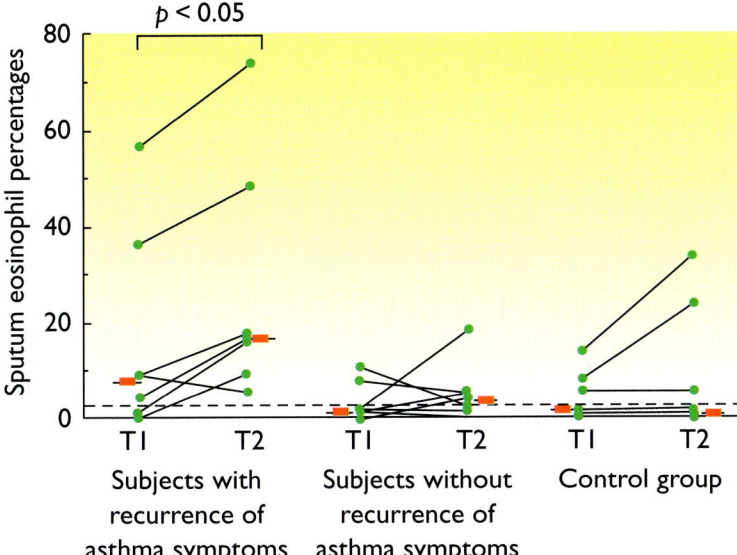

Figure 4.13 Sputum eosinophil counts in well controlled asthmatic patients before (T1) and after a double-blind placebo-controlled withdrawal of inhaled corticosteroids. The patients were followed for 3 months or until an exacerbation occurred (T2). The control group comprised patients who were kept on inhaled steroids. Bars represent medians and the dashed line shows the upper limit of normal values. Reproduced with permission from Giannini D, Di Franco A, Paggiaro PL, *et al.* Analysis of induced sputum before and after withdrawal of treatment with inhaled corticosteroids in asthmatic patients. *Clin Exp Allergy* 2000;30:1777–84

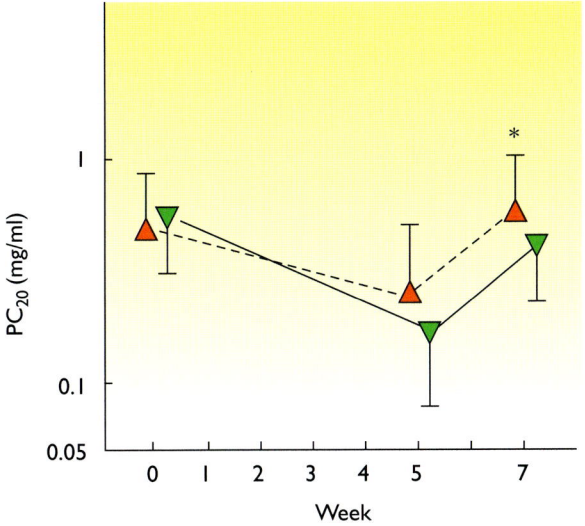

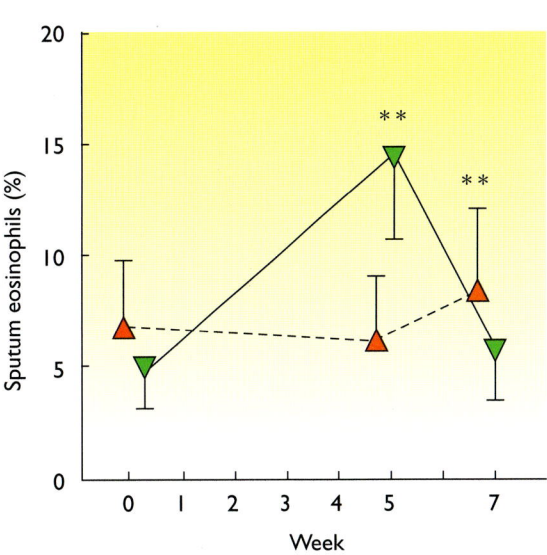

Figure 4.14 Effect of double-blind cross-over tapering of inhaled corticosteroids on airway responsiveness to methacholine (left panel) and sputum eosinophil counts (right panel) in patients with difficult-to-control asthma. Data were collected at run-in (week 0), during exacerbation (week 5), and in remission after receiving treatment with high doses of fluticasone (1000–3000 μg). Results are expressed as mean ± standard error of the mean. The continuous line represents the period during which treatment with fluticasone was tapered, and the dashed line represents the control treatment period during which the dose of inhaled steroid remained constant. *$p < 0.05$; **$p < 0.01$ when comparing values between exacerbation and baseline and between remission and exacerbation within the two groups. Reproduced with permission from in't Veen JC, Smits HH, Bel EH, *et al.* Lung function and sputum characteristics of patients with severe asthma during an induced exacerbation by double-blind steroid withdrawal. *Am J Respir Crit Care Med* 1999;160:93–9

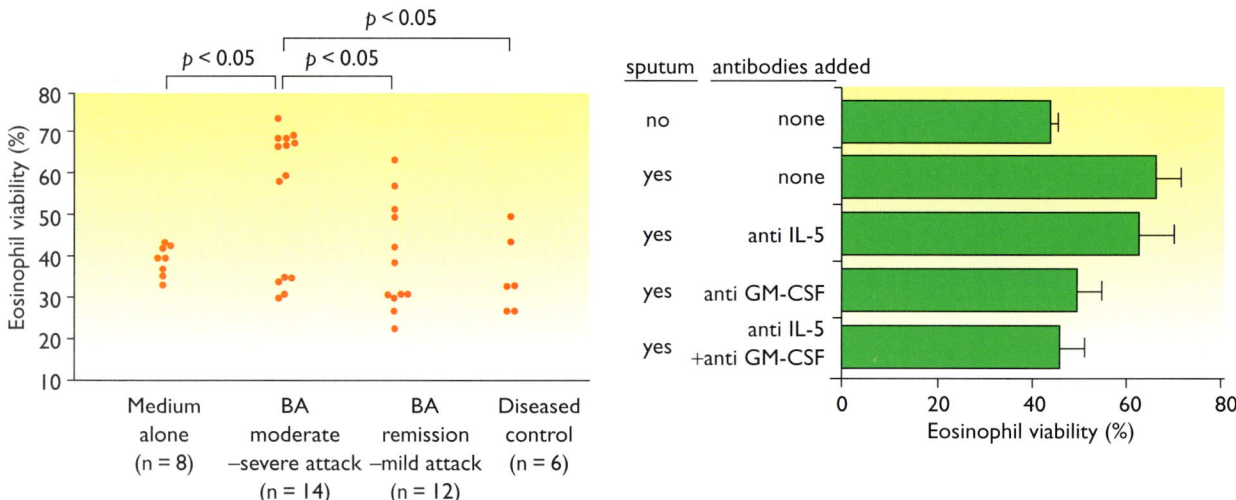

Figure 4.15 Increased eosinophil viability-enhancing activity (EVEA) in sputum from asthmatics during an asthma attack (left). The EVEA was assessed using eosinophils from guinea pigs cultured for 4 days with human sputum supernatants. Seven sputum extracts with high EVEA were pretreated with anti-IL-5 and anti-GM-CSF antibodies for 1 h at 4°C, and then incubated with eosinophils. BA, bronchial asthma. Reproduced with permission from Adachi T, Motojima S, Hirata A, *et al*. Eosinophil viability-enhancing activity in sputum from patients with bronchial asthma. Contributions of interleukin-5 and granulocyte/macrophage colony-stimulating factor. *Am J Respir Crit Care Med* 1995;151:618–23

(a)

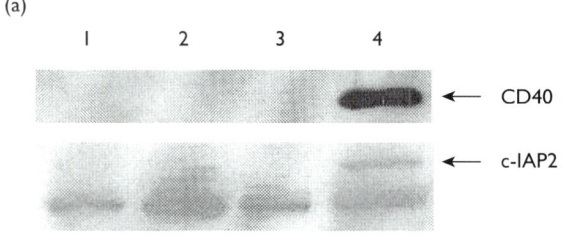

(b)

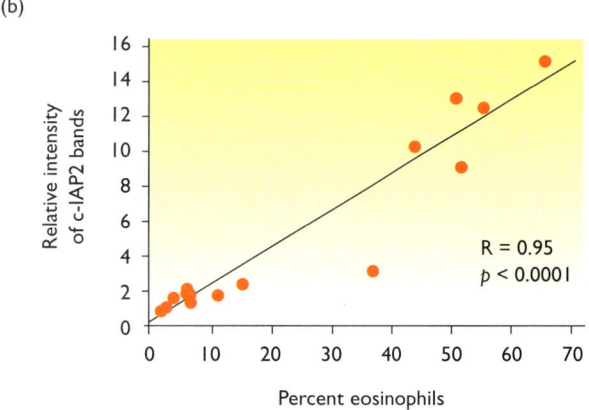

Figure 4.16 Expression of the costimulatory molecule CD40, and the cellular inhibitor apoptosis protein-2 (c-IAP2) in sputum eosinophils from atopic asthmatics. (a) blood eosinophils and sputum cells from healthy subjects (lane 1 and 2 respectively) and atopic asthma patients (lane 3 and 4 respectively) were isolated and assayed for CD40 and c-IAP2 expression by immunoblotting. These show that blood eosinophils and sputum cells from the healthy subject do not express CD40 and c-IAP2. While blood eosinophils from the atopic asthmatic did not express CD40 and c-IAP2, those proteins were clearly seen in sputum cells. There was a correlation between the levels of c-IAP2 expression in sputum cells from asthmatic patients (n = 15), as assessed by photodensitometry, and sputum relative eosinophil counts (correlation assessed by standard least-square linear regressions; R, coefficient of correlation). Reproduced with permission from Bureau F, Seumois G, Lekeux P, *et al*. CD40 engagement enhances eosinophil survival through induction of cellular inhibitor of apoptosis protein 2 expression: possible involvement in allergic inflammation. *J Allergy Clin Immunol* 2002;110:443–9

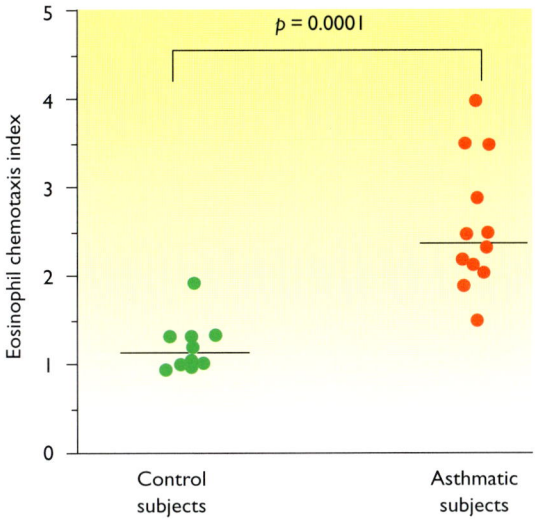

Figure 4.17 Eosinophil chemotactic activity measured in sputum samples from healthy subjects and asthmatics assessed in 48-well microchemotaxis Boyden chambers. The eosinophil chemotactic index is the ratio between the number of cells migrating from upper wells, in response to chemoattractants present in the sputum supernatant placed in lower wells, and those migrating in response to control medium alone. An index of 1, therefore, indicates that induced chemotaxis is the same as spontaneous migration. The bars represent median values. Reproduced with permission from Louis R, Shute J, Djukanovic R, *et al.* Cell infiltration, ICAM-1 expression, and eosinophil chemotactic activity in asthmatic sputum. *Am J Respir Crit Care Med* 1997;155:466–72

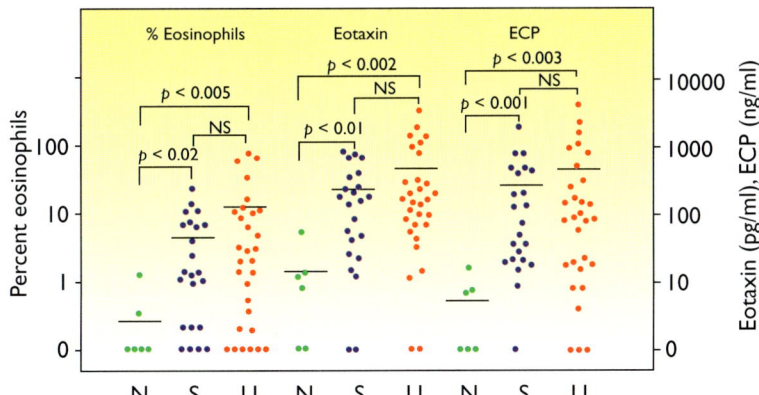

Figure 4.18 Sputum eosinophil percentages and eotaxin and ECP levels in healthy subjects and asthmatics classified as stable or unstable based on symptoms, peak flow variability and β_2-agonist consumption. Horizontal bars show mean values. N, normal; S, stable; U, unstable. Reproduced with permission from Yamada H, Yamaguchi M, Yamamoto K, *et al.* Eotaxin in induced sputum of asthmatics: relationship with eosinophils and eosinophil cationic protein in sputum. *Allergy* 2000;55:392–7

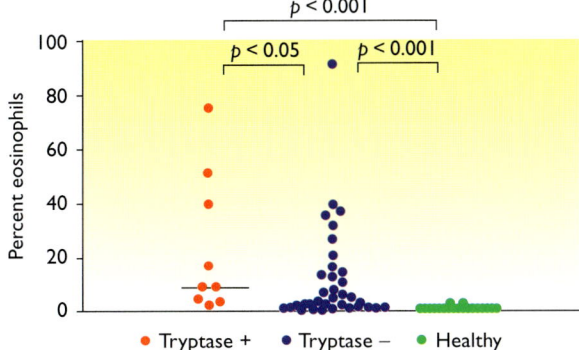

Figure 4.19 The relationship between mast-cell activation, as assessed by sputum tryptase levels, and sputum eosinophilia. Sputum eosinophil counts are shown in asthmatics subdivided according to their sputum tryptase levels. Tryptase+ are those with detectable tryptase i.e. > 1 ng/ml. Horizontal bars represent median values. Reproduced with permission from Bettiol J, Radermecker M, Louis R, *et al.* Airway mast-cell activation in asthmatics is associated with selective sputum eosinophilia. *Allergy* 1999;54:1188–93[17]

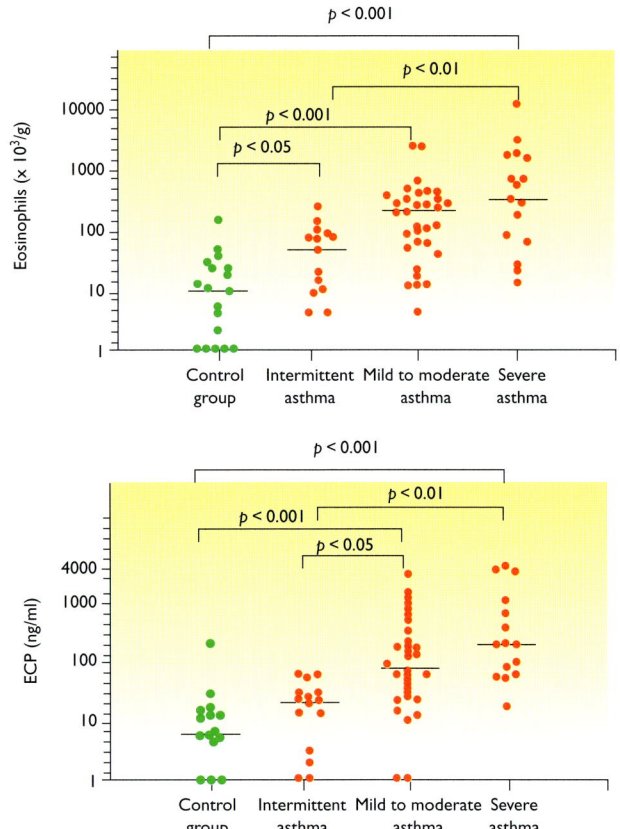

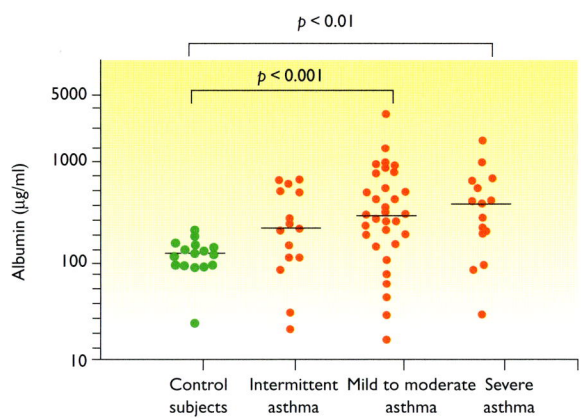

Figure 4.20 The relationship between disease severity and airway inflammation. Sputum eosinophil numbers, ECP and albumin are seen to increase in relation to disease severity (classified according the GINA criteria irrespective of their current treatment), although there is considerable variability within patient categories. Horizontal bars indicate median values. Reproduced with permission from Louis R, Lau LC, Djukanovic R, *et al*. The relationship between airways inflammation and asthma severity. *Am J Respir Crit Care Med* 2000;161:9–16

anti-IL-5 antibodies were given to mild asthmatics and the effects of this on allergen-induced bronchoconstriction and blood and sputum eosinophils studied (Figure 4.24). Despite a dramatic reduction in the number of blood and sputum eosinophils, the treatment was unable either to affect airway responsiveness or prevent the occurrence of the late-phase response that followed allergen challenge. This study has raised the question of the role of the eosinophil as a pivotal cell in asthma. This question notwithstanding, assessment of sputum eosinophilia may help the clinician not only to make a diagnosis but also to develop a therapeutic strategy. When constructing receiver operating characteristics (ROC) curves, sputum eosinophil counts are found to be of greater value than blood eosinophil counts in differentiating asthmatics from healthy subjects and patients with chronic bronchitis (Figure 4.25). Furthermore, when comparing several diagnostic tests for mild to moderate asthma, sputum eosinophil counts are superior than PEF variability and reversibility to β_2-agonists (Figure 4.26). High eosinophil counts are predictive of a good clinical response to inhaled steroids, and some studies show that treatment with these drugs fails to improve control in asthmatics with sputum eosinophil count less than 3% (Figure 4.27).

Given that the proportion of steroid-free, mild to moderate asthmatics with normal eosinophil counts (i.e. ≤ 2% of nonsquamous inflammatory cells) may be as high as 31% (Figure 4.28), it will be important to elucidate to what extent low numbers of eosinophils contribute to asthma and whether there are other cell types that are more important in these patients. The persistence of high sputum eosinophil counts despite good clinical control in patients already treated with inhaled steroids is predictive of

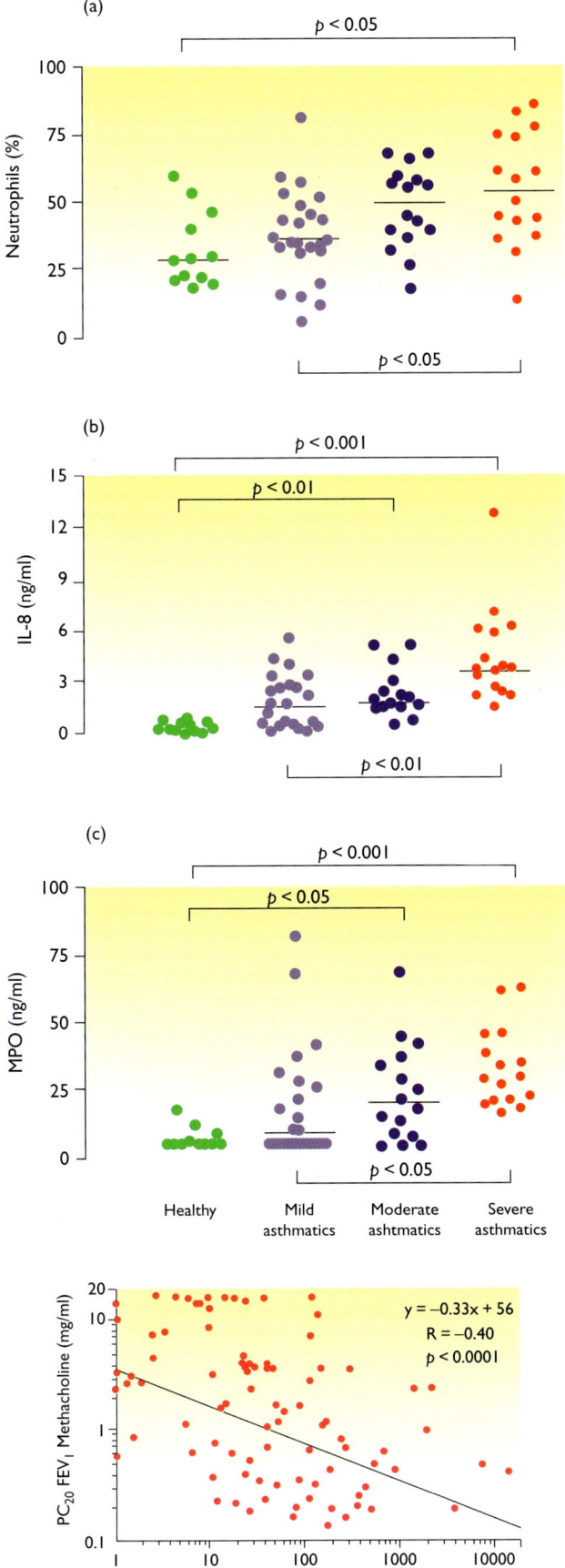

Figure 4.21 The relationship between asthma severity and neutrophilic airway inflammation. Sputum neutrophil percentages, IL-8 and MPO in healthy subjects and 55 asthmatics are shown. Asthmatics were classified as mild, moderate and severe according to the GINA criteria. Horizontal bars represent median values. Reproduced with permission from Jatakanon A, Uasuf C, Barnes PJ, *et al*. Neutrophilic inflammation in severe persistent asthma. *Am J Respir Crit Care* Med 1999;160:1532–9

Figure 4.22 The relationship between sputum eosinophil number and methacholine bronchial responsiveness, assessed by the provocative concentration causing a fall of 20% in FEV_1 (PC_{20}) in 118 mild to moderate steroid-naive asthmatics. R, Pearson's coefficient of correlation. Reproduced with permission from Louis R, Sele J, Bartsch P, *et al*. Sputum eosinophil count in a large population of patients with mild to moderate steroid-naive asthma: distribution and relationship with methacholine bronchial hyperresponsiveness. *Allergy* 2002;57:907–12

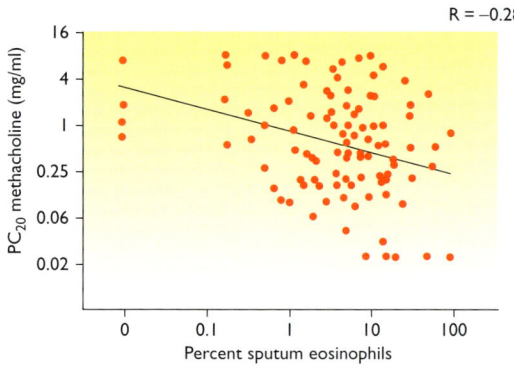

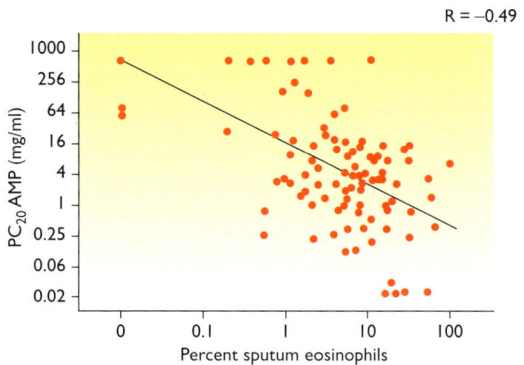

Figure 4.23 The relationship between sputum eosinophils (shown as a percentage of total cells) and methacholine (upper panel) and adenosine (lower panel) bronchial hyperresponsiveness as reflected by the provocative concentration causing a 20% fall in FEV_1 (PC_{20}) in steroid-free asthmatics; R, Pearson's coefficient of correlation. Reproduced with permission from Van Den Berge M, Meijer RJ, Postma DS, *et al.* PC(20) adenosine 5'-monophosphate is more closely associated with airway inflammation in asthma than PC(20) methacholine. *Am J Respir Crit Care Med* 2001;163:1546–50

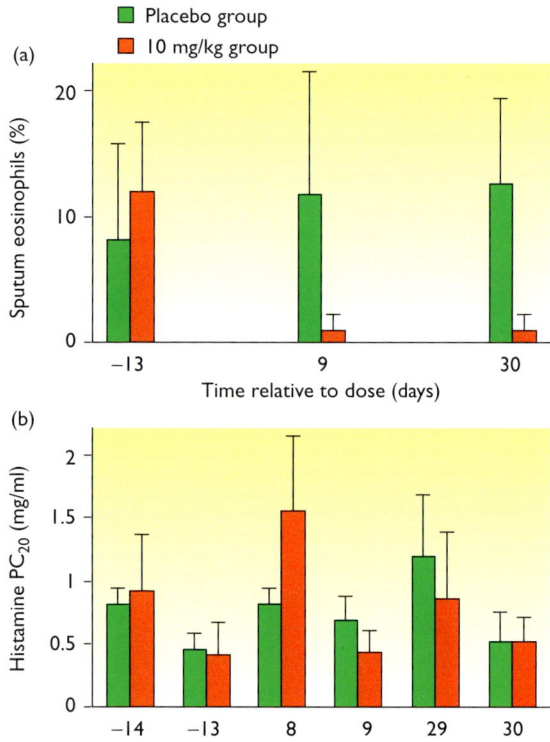

Figure 4.24 The effects of humanized monoclonal antibody to IL-5 on sputum eosinophils and histamine PC_{20}. Subjects underwent baseline histamine PC_{20} measurement pre (-14 days) and post allergen challenge (-13 days) and baseline sputum induction 24 h post allergen challenge (-13 days). This was followed by a single injection of anti-IL-5 antibody. The PC_{20} measurements were repeated pre and post allergen challenge on days 8 and 9, respectively, and once again on days 29 and 30, respectively. Sputum induction was only repeated on days 9 and 30, i.e. 24 h after allergen challenge. (a) the significant effects on eosinophil counts (presented as a percentage of total leukocytes) caused by anti-IL-5 treatment; (b) there was no protective effect on histamine PC_{20}. Data are expressed as arithmetic mean with 95% confidence interval for sputum eosinophils and geometric mean with geometric standard error for histamine PC_{20}. Reproduced with permission from Leckie MJ, ten Brinke A, Barnes PJ, *et al.* Effects of an interleukin-5 blocking monoclonal antibody on eosinophils, airway hyperresponsiveness, and the late asthmatic response. *Lancet* 2000;356:2144–8

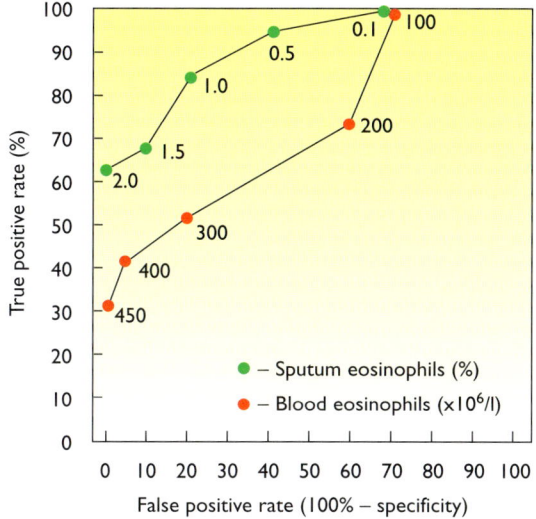

Figure 4.25 Receiver operating characteristics (ROC) curves for sputum and blood eosinophils comparing their values for making a diagnosis of asthma. Plots that lie furthest to the 'northwest' represent more accurate values. Each symbol carries the true positive rate and the false positive rate for a given sputum and blood eosinophil count. Reproduced with permission from Pizzichini E, Pizzichini MM, Efthimiadis A, *et al.* Measuring airway inflammation in asthma: eosinophils and eosinophilic cationic protein in induced sputum compared with peripheral blood. *J Allergy Clin Immunol* 1997;99:539–44

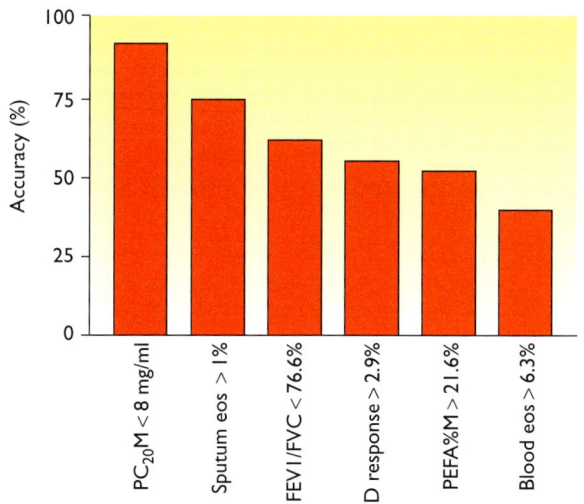

Figure 4.26 Comparison of accuracy of several tests for diagnosing mild asthma. Reproduced with permission from Hunter CJ, Brightling CE, Woltmann G, *et al.* A comparison of the validity of different diagnostic tests in adults with asthma. *Chest* 2002;121:1051–7

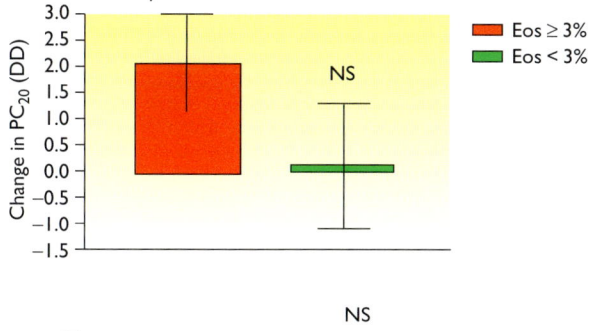

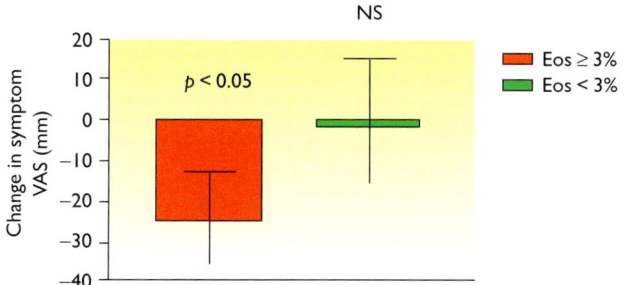

Figure 4.27 Efficacy of inhaled budesonide 800 μg per day administered for 8 weeks on PC_{20} methacholine (upper panel) and symptom scores (lower panel) in asthmatics stratified according to baseline sputum eosinophil counts. No improvement is seen in asthmatics with < 3% eosinophil counts. Results are expressed as mean ± 95% confidence interval. VAS, visual analogue scale; NS, not significant. Reproduced with permission from Pavord ID, Brightling CE, Woltmann G, Wardlaw AJ. Non-eosinophilic corticosteroid unresponsive asthma. *Lancet* 1999;353:2213–14

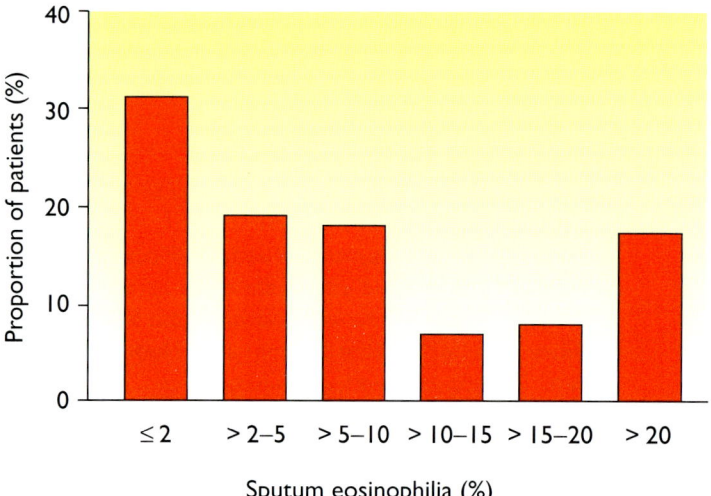

Figure 4.28 Distribution of sputum eosinophil counts (shown as percentage of inflammatory cell counts) in a large group of mild to moderate steroid-naive asthmatics (n = 118) prospectively recruited from an outpatient clinic. Sputum eosinophil percentage ≤ 2% is within the normal range. Reproduced with permission from Louis R, Sele J, Bartsch P, et al. Sputum eosinophil count in a large population of patients with mild to moderate steroid-naive asthma: distribution and relationship with methacholine bronchial hyperresponsiveness. *Allergy* 2002;57:907–12

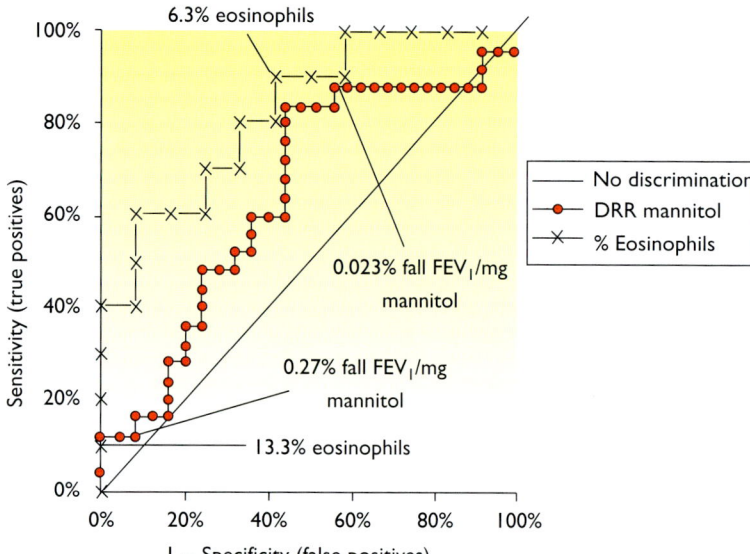

Figure 4.29 Receiver operator characteristic (ROC) curves showing the sensitivity and specificity over a range of cut-points for dose–response ratio mannitol (DRR mannitol), a measure of responsiveness to mannitol, defined as the percentage fall in FEV_1 at the last dose of mannitol, divided by the total dose administered and percent sputum eosinophils for predicting failure to reduce inhaled corticosteroid treatment. The solid line indicates no discrimination. Plots that lie farthest to the 'northwest' indicate more accurate values. Reproduced with permission from Leuppi JD, Salome CM, Woolcock AJ, et al. Predictive markers of asthma exacerbation during stepwise dose reduction of inhaled corticosteroids. *Am J Respir Crit Care Med* 2001;163:406–12

an exacerbation when tapering the doses of inhaled steroids. Leuppi and co-workers found that a sputum eosinophil count of 6.3% was 90% sensitive at predicting the failure of reducing the dose of inhaled corticoids, with 90% specificity at 13.3% (Figure 4.29). Furthermore, sputum eosinophils were more useful in predicting an exacerbation than spirometry, asthma symptoms and exhaled nitric oxide levels.

In conclusion, the technique of sputum induction has provided much useful knowledge of inflammatory processes in asthma, extending and complementing studies employing fiberoptic bronchoscopy. A major contribution has been the ability to study airway inflammation in previously poorly studied categories of asthmatics such as those patients with chronic severe asthma or those experiencing acute exacerbations.

CHAPTER 5

Induced sputum studies in chronic obstructive pulmonary disease

Piero Maestrelli and Cristina E. Mapp

Induced sputum has been widely used to study the inflammatory mechanisms of chronic obstructive pulmonary disease (COPD). There remains a question as to the extent to which sampling the airways reflects changes in small airways and the parenchyma, sites where the disease processes are most active in COPD. These issues notwithstanding, the application of induced sputum has been very informative, helping to define some unique features of COPD.

It has long been appreciated that smoking is the most important risk factor for COPD; however, it is equally recognized that only a proportion of smokers develop a clinically overt disease. The mechanisms of the increased susceptibility in these smokers remain unknown. A longitudinal study of smokers has applied sputum induction to examine whether the airway inflammatory process is different in smokers susceptible to COPD development compared with smokers who are 'resistant', i.e. do not develop clinically relevant disease. The percentages of sputum neutrophils were greater in smokers with COPD than in asymptomatic smokers; these correlated with the annual decline in forced expiratory volume (FEV_1) observed over the 15-year follow-up (Figure 5.1). These results strongly suggest that neutrophils are important in the pathogenesis of COPD.

It is likely that neutrophils accumulate in the airway lumen of smoking individuals as a consequence of recruitment from the circulation, a process that involves cell adherence and chemotactic factors which together recruit inflammatory cells into the lungs. Cigarette smoke stimulates alveolar macrophages and possibly epithelial cells to release inflammatory mediators. Examples of such mediators are interleukin-8 (IL-8) and tumor necrosis factor-alpha (TNF-α), whose concentrations are increased in the sputum of asymptomatic smokers and COPD patients (Figures 5.2 and 5.3). These cytokines may, in turn, induce an influx of neutrophils into the lung. TNF-α promotes expression of adhesion molecules, whereas IL-8 is a potent neutrophil chemoattractant and activator of these cells. Increased expression of adhesion molecule CD11b/CD18, the intracellular adhesion molecule-1 (ICAM-1) ligand, has been observed in COPD, and this expression is related to the degree of airway obstruction (Figure 5.4).

An imbalance between cytokines with different properties has been suggested as playing a role in cigarette smoke-induced lung damage. While the pro-inflammatory cytokines IL-8 and TNF-α may be increased in COPD, the anti-inflammatory cytokine IL-10 has been shown to be reduced in sputum of healthy smokers and, to a greater extent, in smokers with COPD (Figure 5.5). These data suggest that there is not a clear qualitative difference in the inflammatory process detected in the airway lumen between COPD and healthy smokers, but rather a quantitative difference. This contrasts with asthma where eosinophils represent

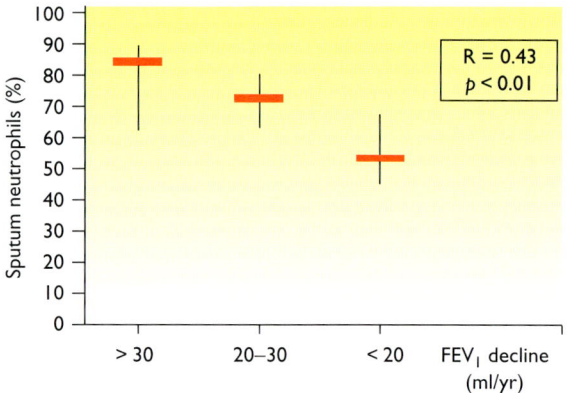

Figure 5.1 Percentage of neutrophils in sputum (median, 25% and 75% percentiles) in subjects with declines in forced expiratory volume (FEV_1) of < 20 ml/year, 20–30 ml/year, and > 30 ml/year. R, Spearman rank correlation coefficient. Reproduced with permission from the BMJ Publishing Group from Stanescu D, Sanna A, Maestrelli P, et al. Airways obstruction, chronic expectoration, and rapid decline of FEV_1 in smokers are associated with increased levels of sputum neutrophils. *Thorax* 1996;51:267–71

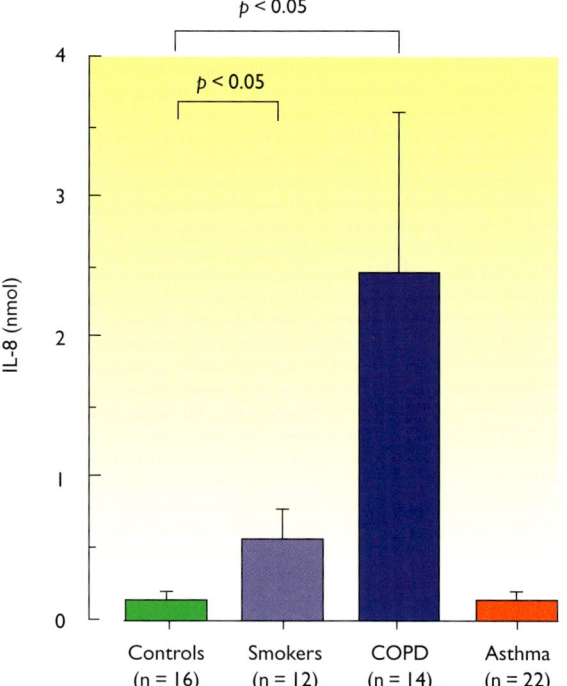

Figure 5.2 Concentrations of interleukin (IL)-8 in the fluid phase of induced sputum from control, smoking, COPD and asthmatic subjects. Mean values ± standard error of the mean are shown. Reproduced with permission from Keatings VM, Collins PD, Scott DM, Barnes PJ. Differences in interleukin-8 and tumor necrosis factor-alpha in induced sputum from patients with chronic obstructive pulmonary disease or asthma. *Am J Respir Crit Care Med* 1996;153:530–4

a prominent inflammatory feature that is virtually absent in normal control subjects (Figure 5.6).

Recent studies have shown considerable overlap between asthma and COPD and the challenge is now to identify distinct features that enable these two conditions to be differentiated. A proportion of patients with COPD exhibit functional features that resemble asthma. These COPD patients show a degree of reversibility of airflow limitation, although the extent of reversibility is not as great as in asthma (Figure 5.7). Some COPD patients also demonstrate hyperresponsiveness to inhaled adenosine 5-monophosphate, a feature that is normally associated with asthma (Figure 5.8). Both these subgroups have more eosinophils in the induced sputum than the COPD sufferers without any clinical or pathophysiological features of asthma.

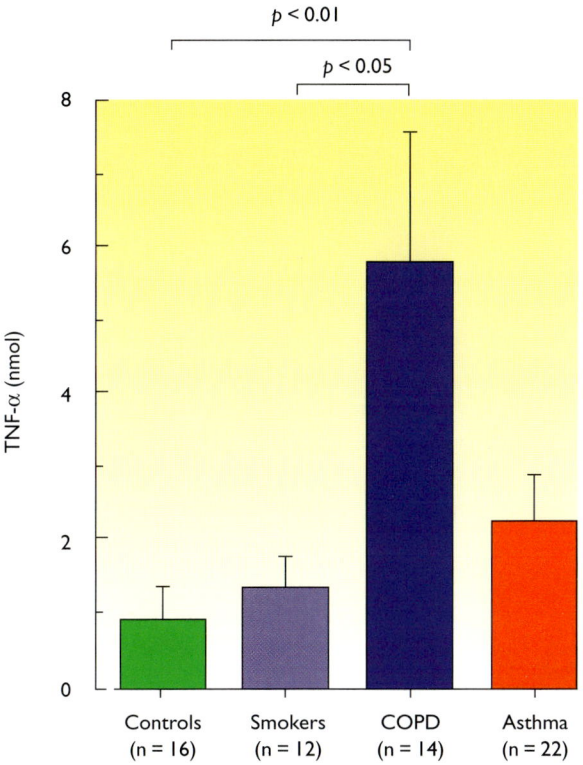

Figure 5.3 Concentrations of tumor necrosis factor (TNF)-α in the fluid phase of induced sputum from control, smoking, COPD and asthmatic subjects. Mean values ± standard error of the mean are shown. Reproduced with permission from Keatings VM, Collins PD, Scott DM, Barnes PJ. Differences in interleukin-8 and tumor necrosis factor-alpha in induced sputum from patients with chronic obstructive pulmonary disease or asthma. *Am J Respir Crit Care Med* 1996;153:530–4

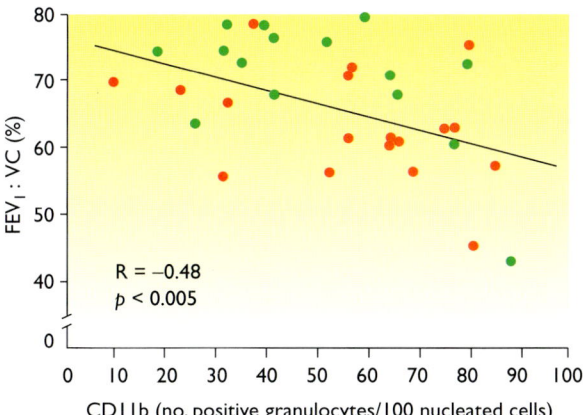

Figure 5.4 Relationship between FEV_1 : VC ratio and expression of CDIIb on granulocytes in subjects with smoking history. The Spearman rank correlation coefficient (R) was calculated for all 33 subjects. Reproduced with permission from Maestrelli P, Calcagni PG, Stanescu D, *et al.* Integrin upregulation on sputum neutrophils in smokers with chronic airway obstruction. *Am J Respir Crit Care Med* 1996;154:1296–300

In most COPD patients, corticosteroids are ineffective at modifying the airway pathology that is more commonly characterized by neutrophilic airway inflammation and detectable by examination of induced sputum. However, eosinophilic airway inflammation, exhibited by some COPD patients, is associated with a significant response to short-term effects of corticosteroid treatment as measured by changes in post-bronchodilator FEV_1, health status and symptom scores. Therefore, sputum eosinophilia may predict a beneficial response to steroid treatment (Figures 5.10 and 5.11). While these observations are interesting, further, larger studies are needed to provide more definitive answers that would be of relevance to routine clinical practice. Other drugs used in the

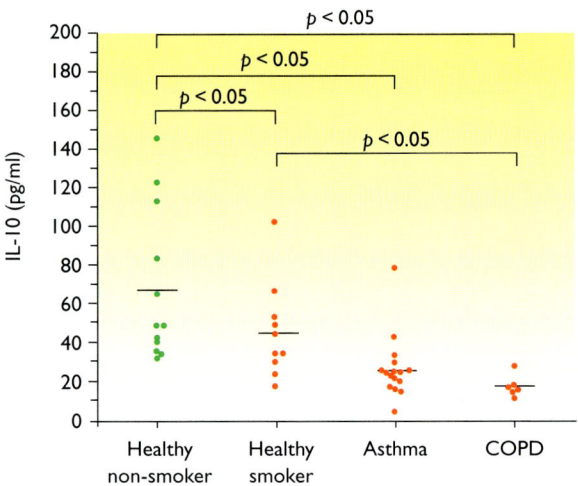

Figure 5.5 Levels of IL-10 in sputum obtained from healthy non-smokers, healthy smokers and patients with asthma and COPD. Horizontal bars represent mean values. Reproduced with permission from Takanashi S, Hasegawa Y, Okamura K, et al. Interleukin-10 level in sputum is reduced in bronchial asthma, COPD and in smokers. Eur Respir J 1999;14:309–14

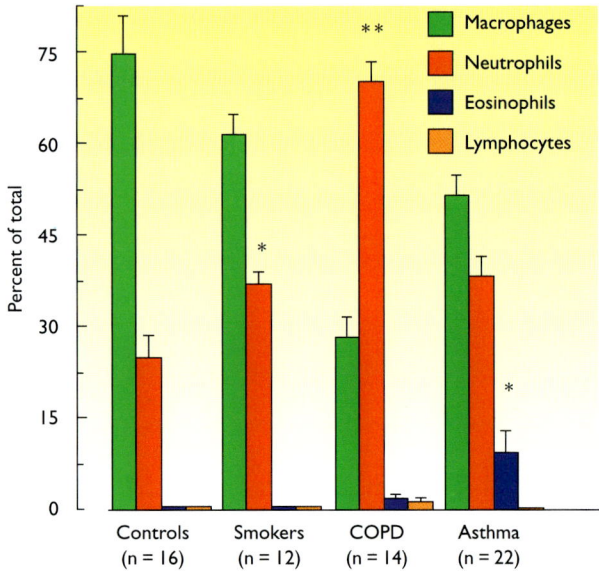

Figure 5.6 Differential cell counts in induced sputum in different subject groups. *$p < 0.05$ versus COPD, smoking, and non-smoking control groups; **$p < 0.001$ versus asthma, smoking, and non-smoking control groups. Reproduced with permission from Keatings VM, Collins PD, Scott DM, Barnes PJ. Differences in interleukin-8 and tumor necrosis factor-alpha in induced sputum from patients with chronic obstructive pulmonary disease or asthma. Am J Respir Crit Care Med 1996;153: 530–4

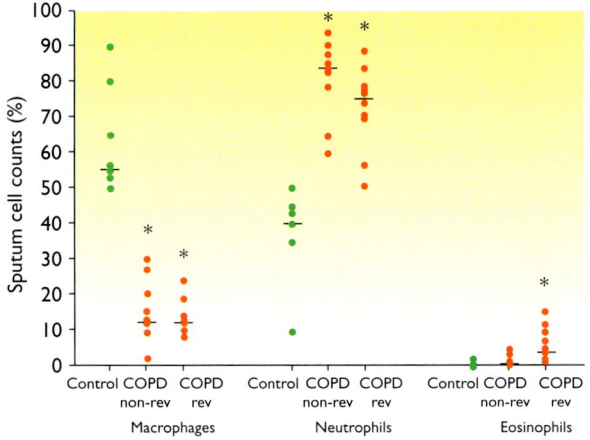

Figure 5.7 Characteristics of sputum inflammatory cells in healthy control subjects and COPD patients with (COPD rev) and without (COPD non-rev) reversibility of airway obstruction, i.e. an increase of $FEV_1 > 200$ ml after 200 µg of salbutamol. Horizontal bars represent median values. *$p < 0.02$ versus control. Reproduced with permission from Papi A, Romagnoli M, Fabbri LM, et al. Partial reversibility of airflow limitation and increased exhaled NO and sputum eosinophilia in chronic obstructive pulmonary disease. Am J Respir Crit Care Med 2000; 162:1773–7

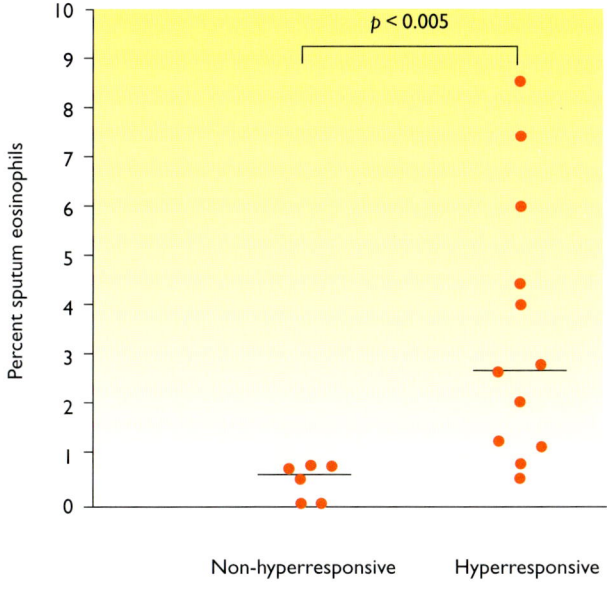

Figure 5.8 Percentage of sputum eosinophils in subjects with COPD with or without hyperresponsiveness to adenosine 5'-monophosphate. Reproduced with permission from Rutgers SR, Timens W, Postma DS, *et al.* Airway inflammation and hyperresponsiveness to adenosine 5'-monophosphate in chronic obstructive pulmonary disease. *Clin Exp Allergy* 2000;30:657–62

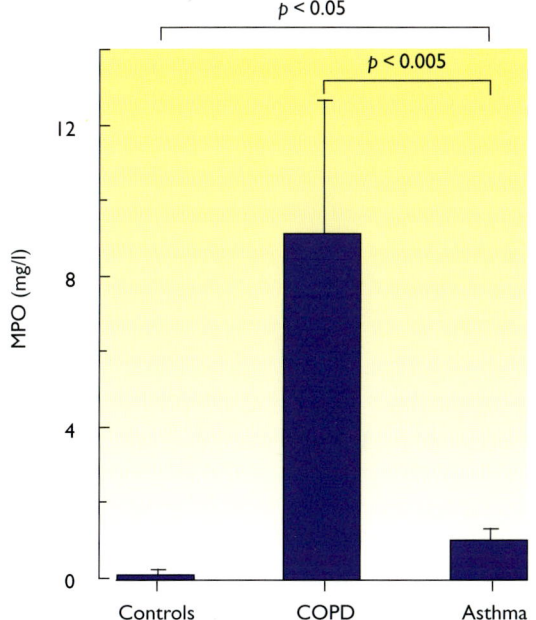
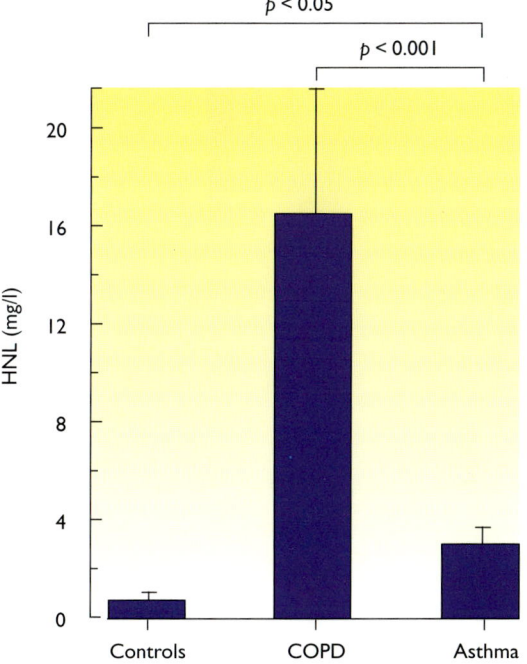

Figure 5.9 Concentrations of myeloperoxidase (MPO) (left panel) and human neutrophil lipocalin (HNL) (right panel) in the subject groups. Reproduced with permission from Keatings VM, Barnes PJ. Granulocyte activation markers in induced sputum: comparison between chronic obstructive pulmonary disease, asthma, and normal subjects. *Am J Respir Crit Care Med* 1997;155:449–53

treatment of COPD have also been shown to be effective at reducing inflammatory indices in COPD: low doses of theophylline have reduced neutrophil counts and IL-8, myeloperoxidase (MPO) and lactoferrin levels in sputum, suggesting that this drug has anti-inflammatory properties

that may be useful in the long-term treatment of COPD (Figure 5.12).

Concentrations of eosinophil cationic protein (ECP) and eosinophil peroxidase (EPO) can be seen to be raised in patients with asthma and those with COPD, despite low numbers of sputum

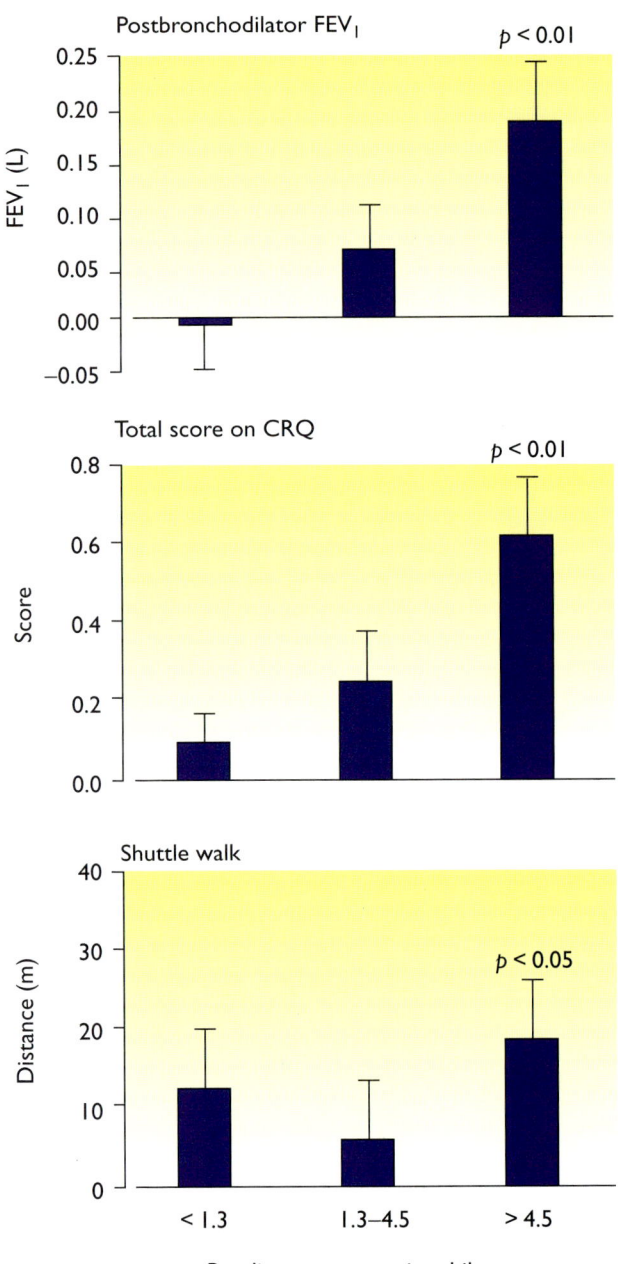

Figure 5.10 Mean (bars represent standard error of the mean) absolute increase in primary outcomes (post-bronchodilator FEV$_1$, chronic respiratory disease questionnaire (CRQ) score, and walking distance) for each tertile of percentages of sputum eosinophils after prednisolone versus placebo. Reproduced with permission from Brightling CE, Monteiro W, Pavord ID, et al. Sputum eosinophilia and short-term response to prednisolone in chronic obstructive pulmonary disease: a randomised controlled trial. *Lancet* 2000;356:1480–5

eosinophils in the latter group. This shows that soluble markers of granulocyte activation in induced sputum are not necessarily related to cell numbers. In contrast, concentrations of the neutrophil products MPO and neutrophil lipocalin (HNL) are significantly higher in patients with COPD than those with asthma, suggesting that these products may be better markers for discriminating between asthma and COPD (Figure 5.9).

The natural history of COPD includes exacerbations that are defined on clinical grounds by increased dyspnea, cough, and sputum production. The origin of exacerbations is complex and not well understood. Exacerbations may result from viral or bacterial airway infection or both. However, they may also occur without demonstrable infection. No consistent differences in sputum total and differential cell counts have been described between stable

p < 0.01
Dyspnea domain score
7.0
6.0
5.0
4.0
3.0
2.0
1.0
p < 0.01
Quality of life score
25
20
15
10
5.0
p < 0.05
Post bronchodilator FEV1 (L)
2.0
1.5
1.0
0.5
0
p < 0.01
Sputum eosinophils (%)
50
40
30
20
10
0
Ba
Pl
Pr
Ba
Pl
Pr
Ba
Pl
Pr
Ba
Pl
Pr

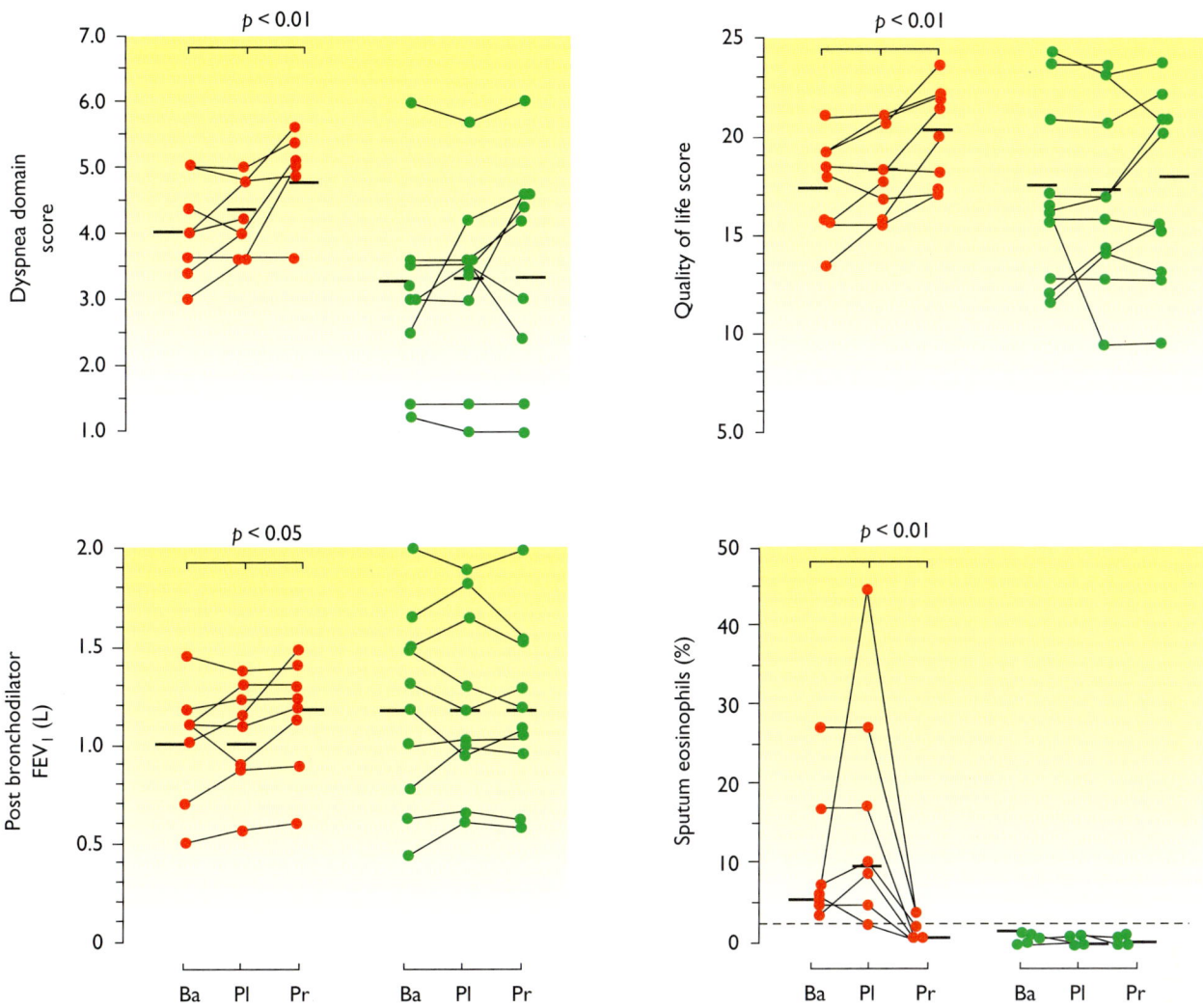

Figure 5.11 Individual values of post-bronchodilator FEV_1, dyspnea domain, quality of life, and sputum eosinophils. Red circles are values from subjects with sputum eosinophilia (eosinophils ≥ 3%) and green circles represent values from subjects without sputum eosinophilia. Horizontal bars represent median values for sputum eosinophils. The dashed line represents the upper limit of normal range. Prednisolone treatment improved clinical outcomes only in subjects with sputum eosinophilia; $^{*}p = 0.05$; $^{**}p < 0.01$; Ba, Baseline; Pl, Placebo; Pr, Prednisone. Reproduced with permission from Pizzichini E, Pizzichini MM, Hargreave FE, *et al.* Sputum eosinophilia predicts benefit from prednisone in smokers with chronic obstructive bronchitis. *Am J Respir Crit Care Med* 1998;158:1511–17

and exacerbated COPD, but the acute response is documented by an increase in markers of airway neutrophilic inflammation. Indeed, concentrations of TNF-α and IL-8 are increased in the sputum of COPD patients during exacerbation compared with when they are clinically stable, and decline significantly 1 month after exacerbation (Figure 5.13).

The number of exacerbations has been found to correlate with levels of sputum IL-6 and IL-8 measured during the stable phase of COPD (Figure 5.14), suggesting that increased baseline levels of inflammatory cytokines in the airways may predict the frequency of future exacerbations. Since total cell, neutrophil and macrophage counts in sputum are not increased in patients with more frequent exacerbations, the origin of IL-6 and IL-8 remains undetermined. Bronchial epithelial cells might be responsible for the

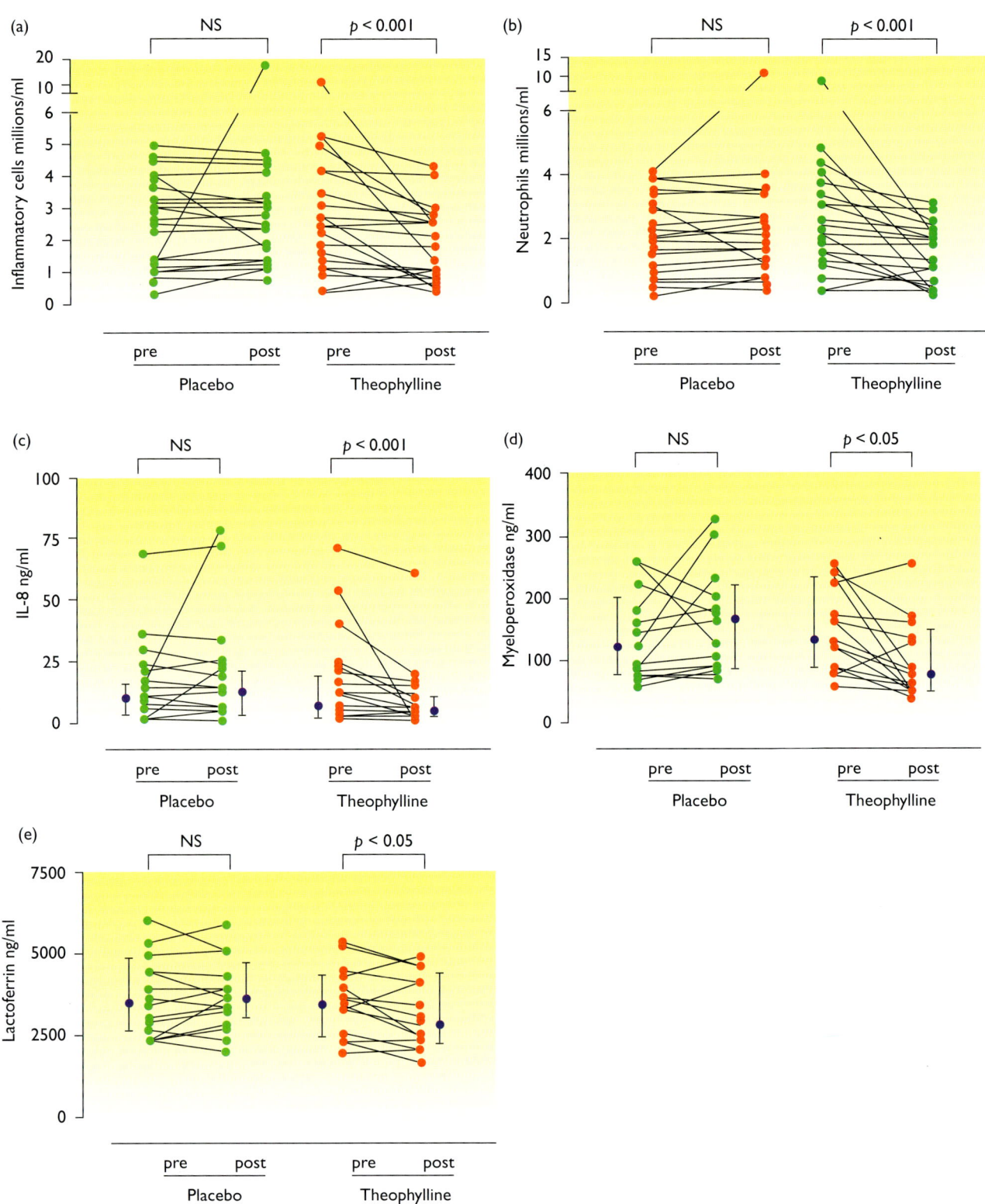

Figure 5.12 Effect of theophylline on induced sputum inflammatory markers and neutrophils in COPD. Data are presented for pre and post placebo and pre and post theophylline treatment. (a) Total cell counts; (b) neutrophil counts; (c) interleukin (IL)-8 levels; (d) myeloperoxidase (MPO) levels; and (e) lactoferrin levels; NS, not significant. Reproduced with permission from Culpitt SV, de Matos C, Barnes PJ, *et al*. Effect of theophylline on induced sputum inflammatory indices and neutrophil chemotaxis in chronic obstructive pulmonary disease. *Am J Respir Crit Care Med* 2002;165:1371–6

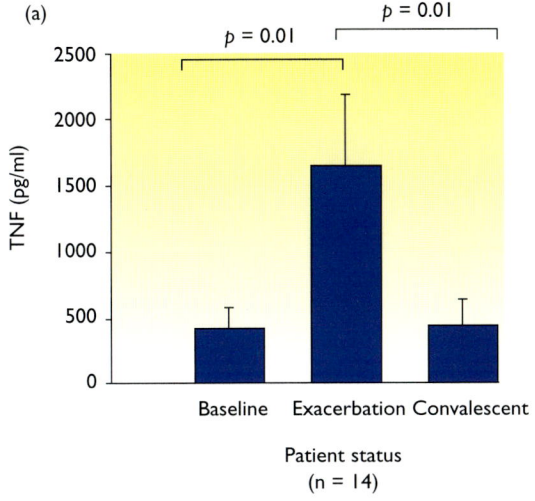

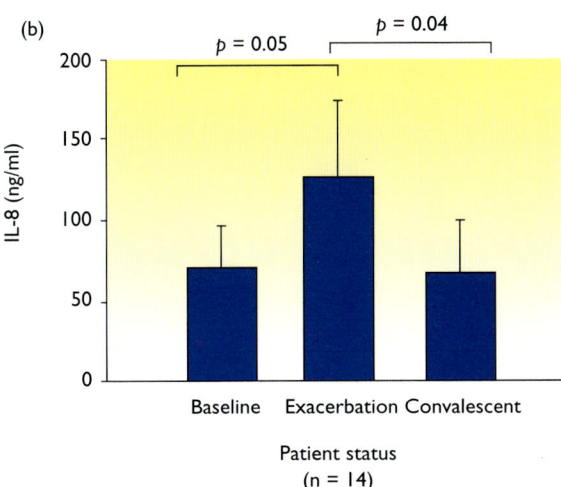

Figure 5.13 Mean sputum tumor necrosis factor (TNF)-α and interleukin (IL)-8 concentrations at the baseline visit, during the exacerbation, and 1 month later. TNF-α and IL-8 concentrations in the fluid phase of sputum were significantly higher at the time of exacerbation visits compared with levels at baseline, and they declined significantly from exacerbation levels when measured 1 month later. Reproduced with permission from Aaron SD, Angel JB, Dales RE, *et al.* Granulocyte inflammatory markers and airway infection during acute exacerbation of chronic obstructive pulmonary disease. *Am J Respir Crit Care Med* 2001;163:349–55

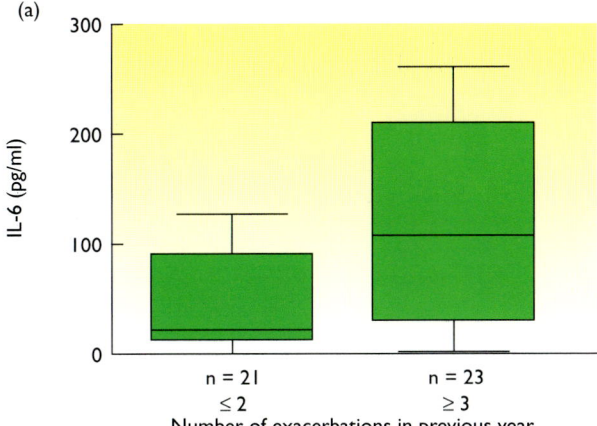

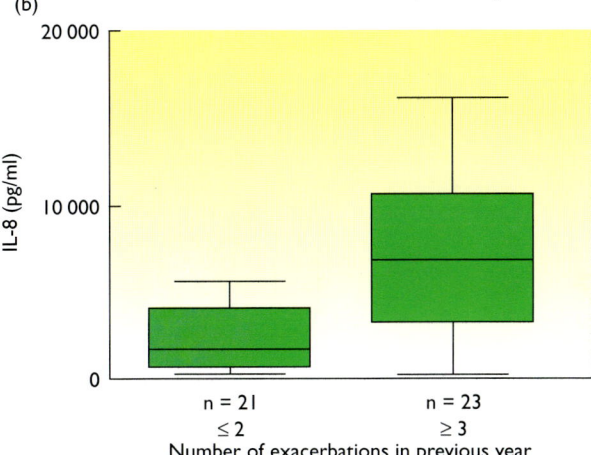

Figure 5.14 Induced sputum levels of (a) interleukin (IL)-6 and (b) IL-8 in patients categorized as frequent exacerbators (≥ 3 exacerbations in the previous year) and infrequent exacerbators (≤ 2 exacerbations in the previous year). Frequent exacerbators exhibit higher levels of sputum IL-6 ($p < 0.05$) and IL-8 ($p < 0.05$). Data expressed as medians (IQR). Adapted with permission from the BMJ Publishing Group from Bhowmik A, Seemungal TA, Sapsford RJ, Wedzicha JA. Relation of sputum inflammatory markers to symptoms and lung function changes in COPD exacerbations. *Thorax* 2000;55:114–20

increased production of these cytokines. Further studies are required to investigate whether frequent exacerbations may increase airway inflammation and thus contribute to the decline in lung function.

In conclusion, analysis of induced sputum has been shown to be useful in the characterization of the type and extent of airway inflammation in COPD, and in defining variants of the disease with overlapping features of asthma. Induced sputum may be useful to determine the effects of new treatments on lung secretions and the relationship between the concentrations of airway cytokines and the course (and possibly prevention) of COPD exacerbations.

CHAPTER 6

Induced sputum studies in children

Peter G. Gibson, Peter Wark and Jodie L. Simpson

Airway inflammation is a major characteristic of childhood asthma. Investigation of the mechanisms and features of airway inflammation may be of value in the diagnosis and clinical management of pediatric airway disease. It is increasingly recognized that objective markers are required in order to monitor the benefits of treatment. Since inflammation is fundamental to the pathophysiology of asthma, and is a key treatment target, the best objective markers are likely to be those that measure airway inflammation. Induced sputum is one source of such markers that holds promise for the management of pediatric asthma.

Sputum can be induced from children using inhalation of hypertonic saline, which is delivered via an ultrasonic nebulizer, either as a 4.5% solution or as increasing concentrations of saline (3, 4 and 5%). Children between the ages of 6 and 18 years have been studied by sputum induction. Successful sputum induction below the age of 6 years may be limited by spirometric technique and a low tidal volume, which restricts the dose of saline that can be delivered.

As in adults, inhalation of hypertonic saline may cause airway narrowing in children; this can be reduced by pretreatment with a β_2-agonist, and most studies of sputum induction in children have used this approach. An alternative technique is to combine sputum induction and bronchial provocation challenge using hypertonic saline. This combined challenge provides a measure of both airway inflammation and airway responsiveness in a single test (Figure 6.1). Whichever technique is used, it is necessary to monitor lung function during sputum induction in order to detect and treat airway obstruction.

NORMAL VALUES

The normal range for sputum cell counts in children is now well established (Table 6.1). The dominant cell in sputum from normal children is the macrophage, and the upper normal limit for sputum eosinophils in children is 2.5%.

SPUTUM EOSINOPHILS IN CHILDREN

The eosinophil has an important role in the pathogenesis of asthma. Eosinophils and the biochemical marker eosinophil cationic protein (ECP) are higher in the sputum of asthmatic children than in sputum from healthy children (Table 6.1, Figure 6.2). Sputum eosinophilia may also reflect disease activity in childhood asthma (Figure 6.3); it correlates with objective markers of disease severity in childhood asthma. Higher levels of induced-sputum eosinophils are associated with greater airflow obstruction (forced expiratory volume in 1 second, FEV_1), and more frequent asthma episodes (Figure 6.3). Sputum eosinophilia

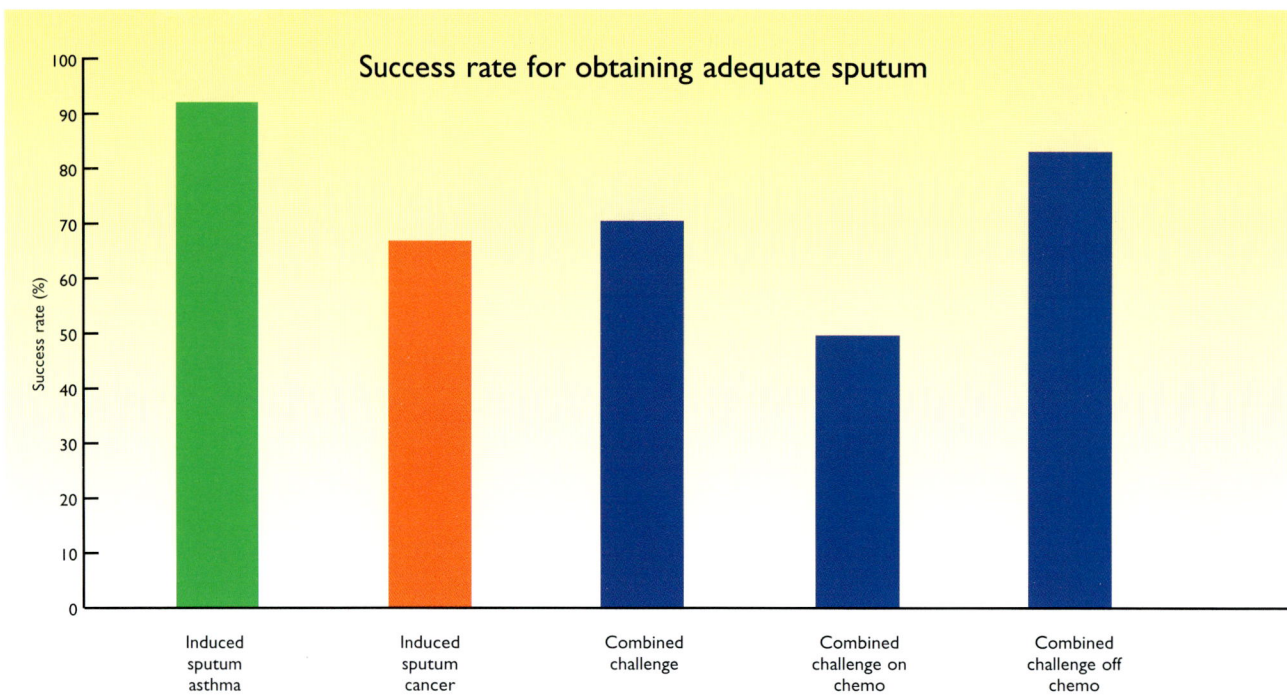

Figure 6.1 Success rate for obtaining an adequate induced sputum sample using sputum induction alone or the combined saline challenge with sputum induction. Reproduced with permission from Jones PD, Hankin R, Simpson J, *et al*. The tolerability, safety and success of sputum induction and combined hypertonic saline challenge in children. *Am J Respir Crit Care Med* 2001;164:1146–9

Table 6.1 Cell counts in sputum from normal children. Reproduced with permission from Cai Y, Carty K, Henry RL, Gibson PG. Persistence of sputum eosinophilia in children with controlled asthma when compared with healthy children. *Eur Respir J* 1998;11:848–53

Sputum	Normal	Atopic normal	Nonatopic normal
TCC x 10^6 cells mL^{-1}			
Mean (95% CI)	5.14 (1.2, 9.08)	1.75 (0.89, 2.6)	8.04 (0.63, 15.5)
Median (IQR)	1.5 (0.8, 3.9)	1.0 (0.55, 2.15)	1.8 (1.05, 6)
Eosinophils % sputum			
Mean (95% CI)	1.57 (0.62, 2.52)	2.16 (0.83, 3.48)	1.13 (0, 2.54)
Median (IQR)	0.3 (0, 1.05)	0.5 (0, 2.8)	0 (0, 0.6)
Mast cells % sputum			
Mean (95% CI)	0.024 (0, 0.05)	0.03 (0, 0.07)	0.02 (0, 0.06)
Median (IQR)	0 (0, 0)	0 (0, 0)	0 (0, 0)

TCC, total cell count; IQR, interquartile range; CI, confidence interval

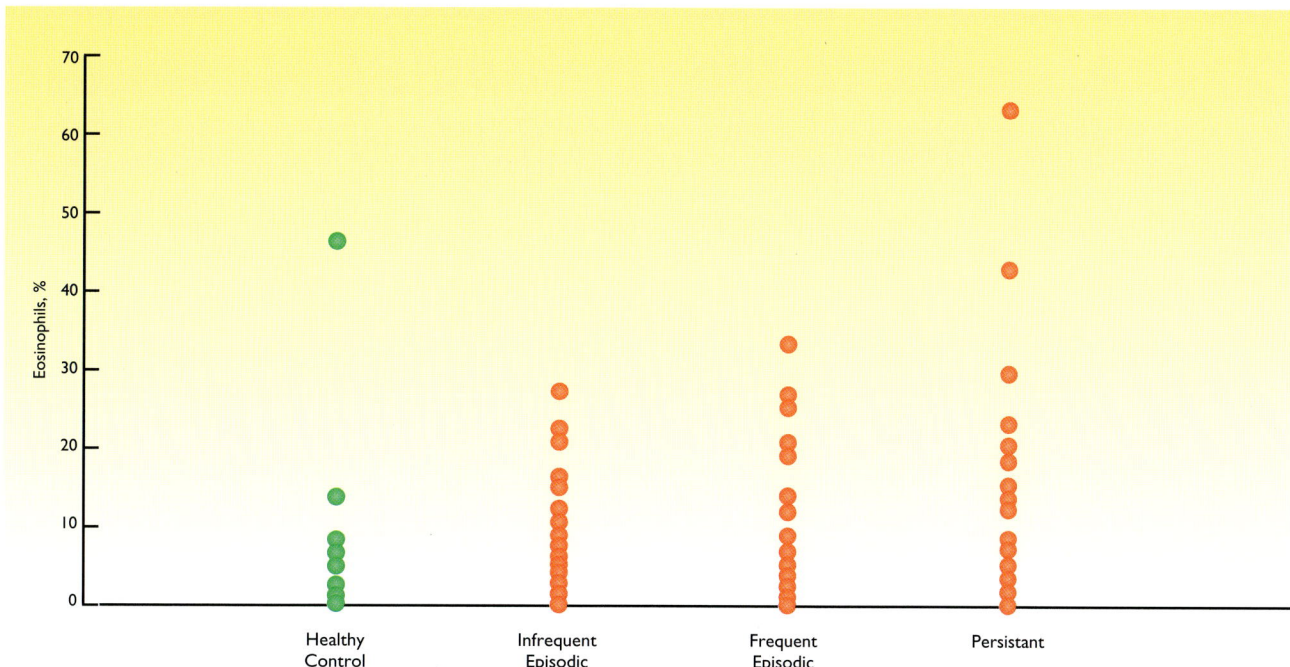

Figure 6.2 Sputum eosinophils in children with infrequent episodic, frequent episodic and persistent asthma. Reproduced with permission from Gibson PG, Simpson JL, Hankin R, *et al*. Induced sputum eosinophils are related to clinical asthma severity in childhood asthma. *Thorax* 2003;58:116–21

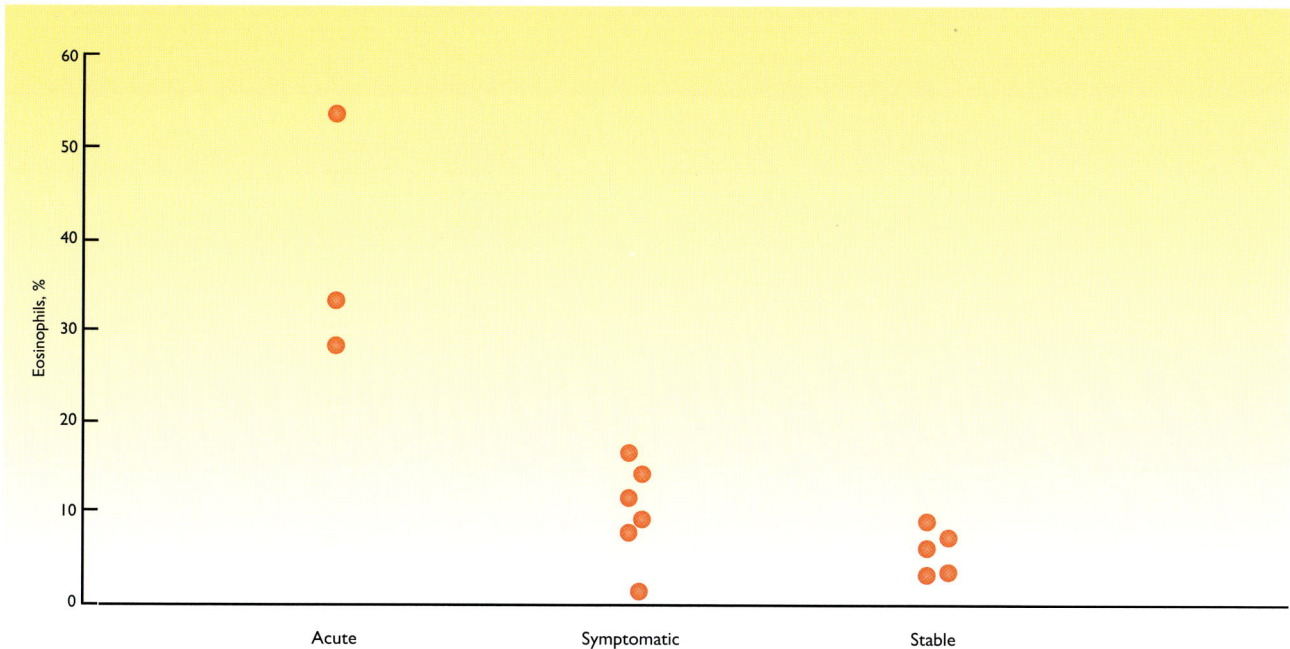

Figure 6.3 Induced-sputum eosinophils in acute, symptomatic, and stable asthma in children. Reproduced with permission from Gibson PG, Simpson JL, Hankin R, *et al*. Induced sputum eosinophils are related to clinical asthma severity in childhood asthma. *Thorax* 2003;58:116–21

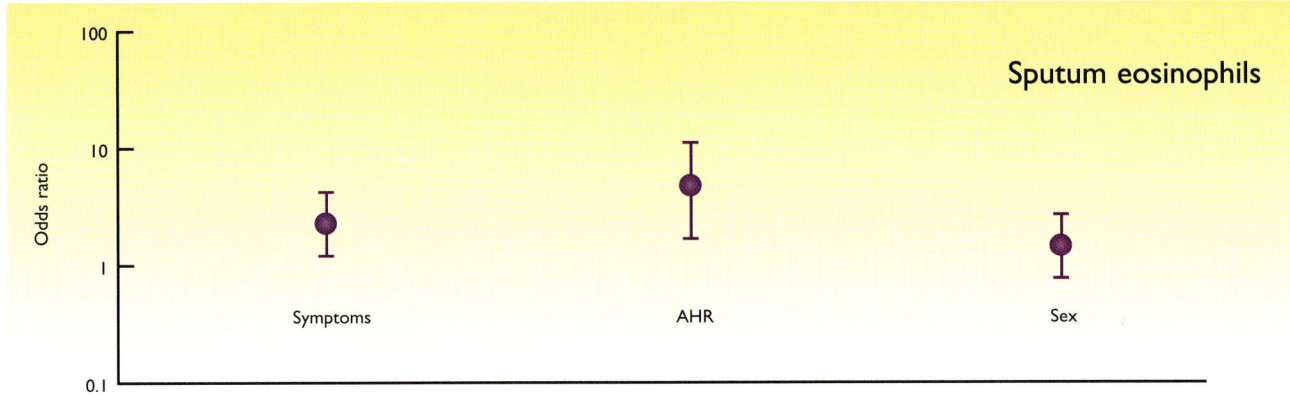

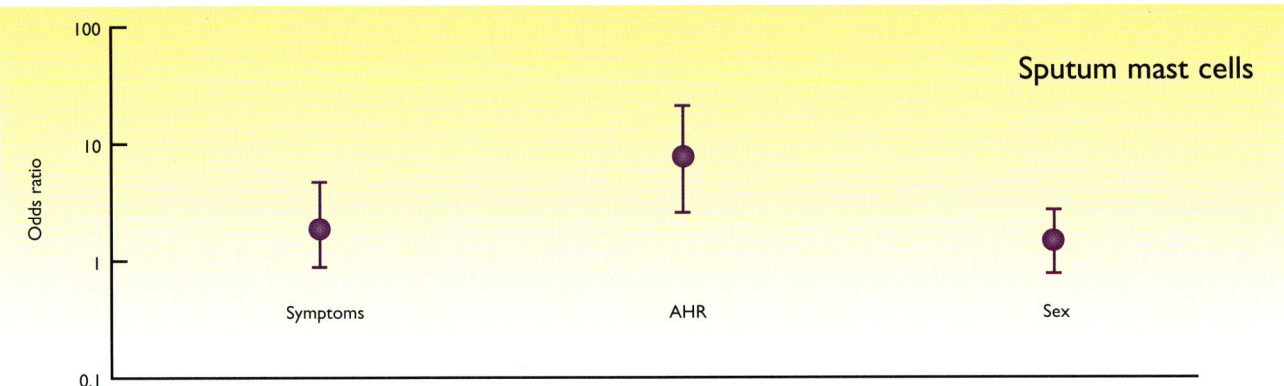

Figure 6.4 Odds ratio (vertical bars show 95% confidence intervals) for the association between sputum eosinophils (top panel), sputum mast cells (bottom panel), and asthma symptoms ($p < 0.05$), airway hyperresponsiveness (AHR) ($p < 0.05$), and sex ($p < 0.05$ for mast cells and male sex). Reproduced with permission from Gibson PG, Wlodarczyk JW, Clancy RL, *et al.* Epidemiological association of airway inflammation with asthma symptoms and airway hyper-responsiveness in childhood. *Am J Respir Crit Care Med* 1998;158:36–41

has also been associated with the degree of airway responsiveness (Figure 6.4).

ASTHMA EXACERBATIONS

Analysis of sputum has been very useful in studying the mechanisms of asthma exacerbations. This has been done using either spontaneously expectorated sputum or sputum induced with normal saline inhalation since administration of hypertonic saline poses a risk of excessive broncho-constriction in such cases. Sputum eosinophilia increases several-fold during exacerbations of asthma. In mild exacerbations with symptomatic deterioration, but no major decline in lung function,

there is a predominant sputum eosinophilia with eosinophil degranulation and ECP release. These markers parallel the changes in asthma symptoms during an exacemrbation. Importantly, sputum markers deteriorate when symptoms exacerbate even though there is no major decline in lung function. This suggests that sputum is a sensitive marker of disease activity in childhood asthma (Figure 6.5).

Exacerbations triggered by spring thunderstorm exposure are thought to be due to massive allergen inhalation. This is accompanied by a marked sputum eosinophilia (Figure 6.6). However, severe exacerbations of asthma causing hospitalization are usually triggered by virus infection, typically rhinovirus. Sputum induction can be used to

(a)
Asthma score
2
1.5
1
0.5
0
Baseline
Day 1
Day 3
Day 7
Day 28
(b)
Percent predicted FEV1
1100
90
80
70
60
50
Baseline
Day 1
Day 3
Day 7
Day 28
(c)
Eosinophils (%)
7
6
5
4
3
2
1
0
Baseline
Day 1
Day 3
Day 7
Day 28
(d)
ECP (ng/ml)
1400
1200
1000
800
600
400
200
0
Baseline
Day 1
Day 3
Day 7
Day 28

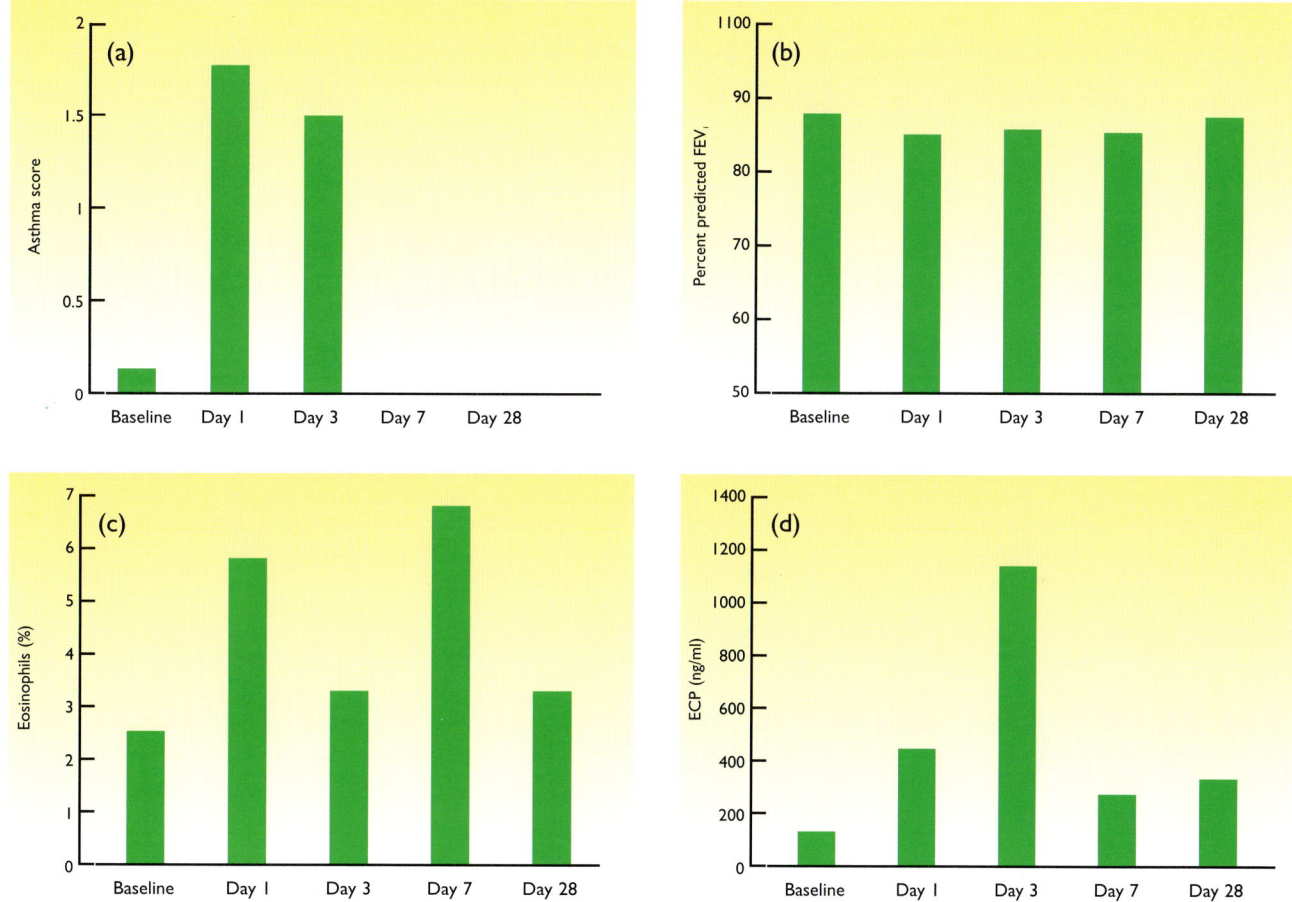

Figure 6.5 Induced sputum in exacerbations of childhood asthma. (a) Asthma symptom score recorded by child at baseline and during exacerbation; (b) lung function represented by FEV_1 at baseline and during exacerbation; (c) sputum eosinophils as a proportion of non-squamous cells at baseline and during exacerbation; and (d) sputum fluid phase eosinophil cationic protein (ECP) at baseline and during exacerbation. Reproduced with permission from Chang AB, Gibson PG, Masters IB, *et al*. The relationship between inflammation and dipalmitoyl phosphatidycholine in induced sputum in children with asthma. *J Asthma* 2003;40:63–70

detect viral infections in asthma using polymerase chain reaction (PCR)-based techniques, and is more sensitive than serology or antigen detection methods (Figures 6.7 and 6.8). The inflammatory cell profile in such severe exacerbations is also different from allergen-induced exacerbations, as there is a prominent neutrophilia rather than eosinophilia (Figure 6.9).

SPUTUM MAST CELLS

Mast cells are seldom seen in the sputum of healthy children (Table 6.1). They may be infrequently seen in sputum from children with asthma, where they are correlated with airway responsiveness to 4.5% saline (Figure 6.4).

COUGH

Chronic cough is another common problem in children. Sputum eosinophilia can occur in this condition; as in adults this is termed eosinophilic bronchitis, and it occurs in about 20% of children with isolated chronic cough without any other features of asthma (Figure 6.10).

In conclusion, non-invasive markers in induced sputum are an important measure of airway

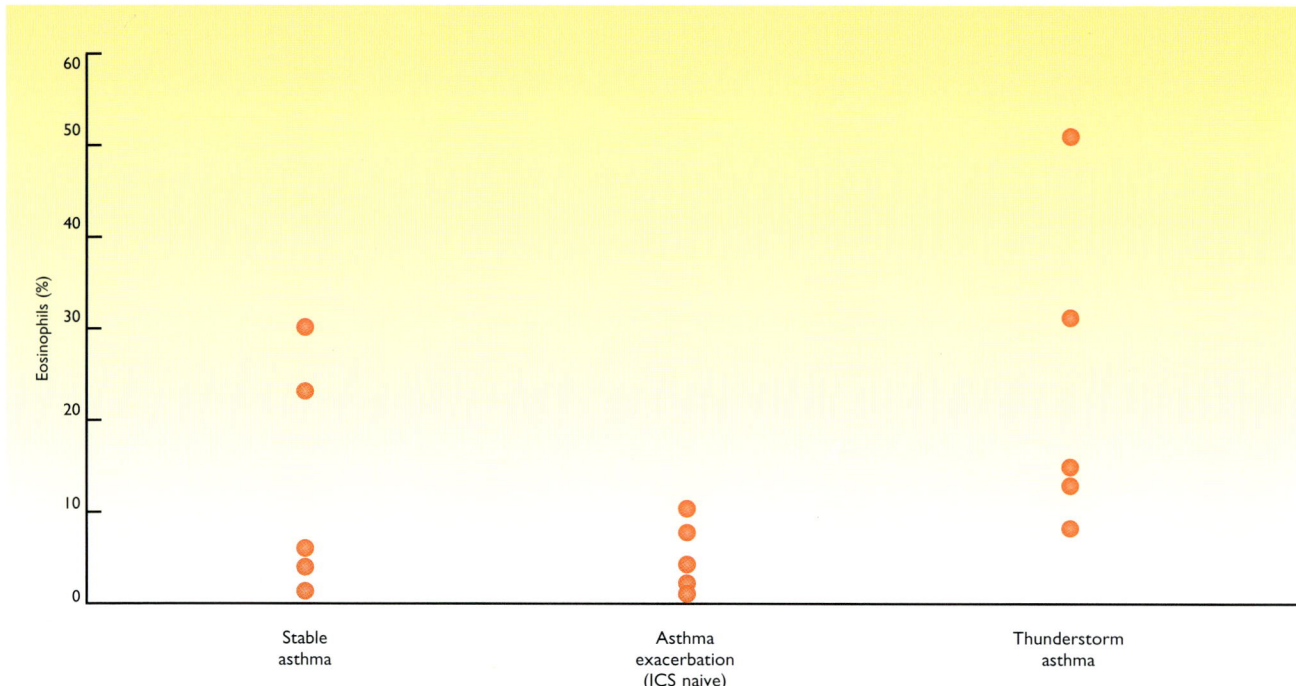

Figure 6.6 Sputum eosinophils in subjects with thunderstorm asthma and two control groups. Reproduced with permission from Wark PAB, Simpson J, Hensley MJ, Gibson PG. Airways inflammation in thunderstorm asthma. *Clin Exp Allergy* 2002;32:1750–6

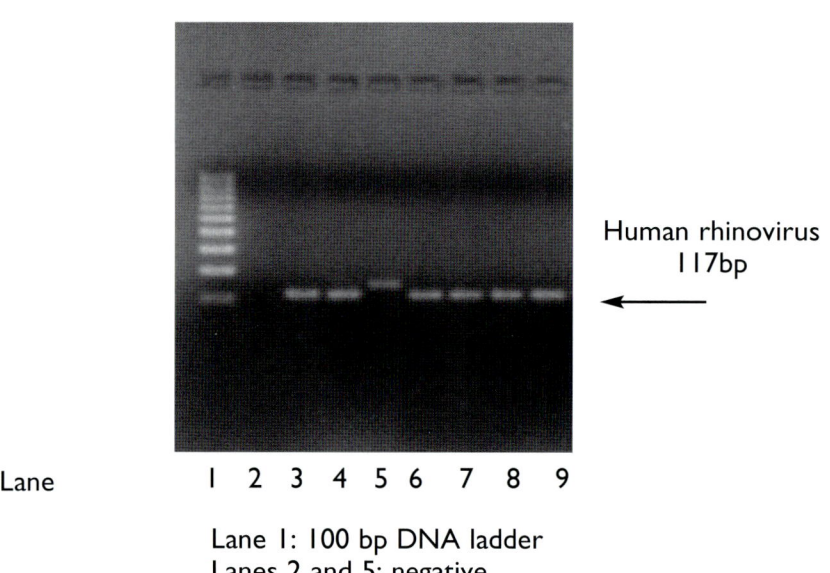

Human rhinovirus
117bp

Lane   1   2   3   4   5   6   7   8   9

Lane 1: 100 bp DNA ladder
Lanes 2 and 5: negative
Lane 9: positive control

Figure 6.7 An agarose gel of induced sputum samples from an asthma exacerbation. The PCR is positive for human rhinovirus in samples shown in lanes 3, 4 and 6–8. Reproduced with permission from Simpson JL, Moric I, Wark PAB, *et al.* Use of induced sputum for the diagnosis of influenza and respiratory syncytial virus infections in asthma: a comparison of diagnostic techniques. *J Clin Virol* 2003;26:339–46

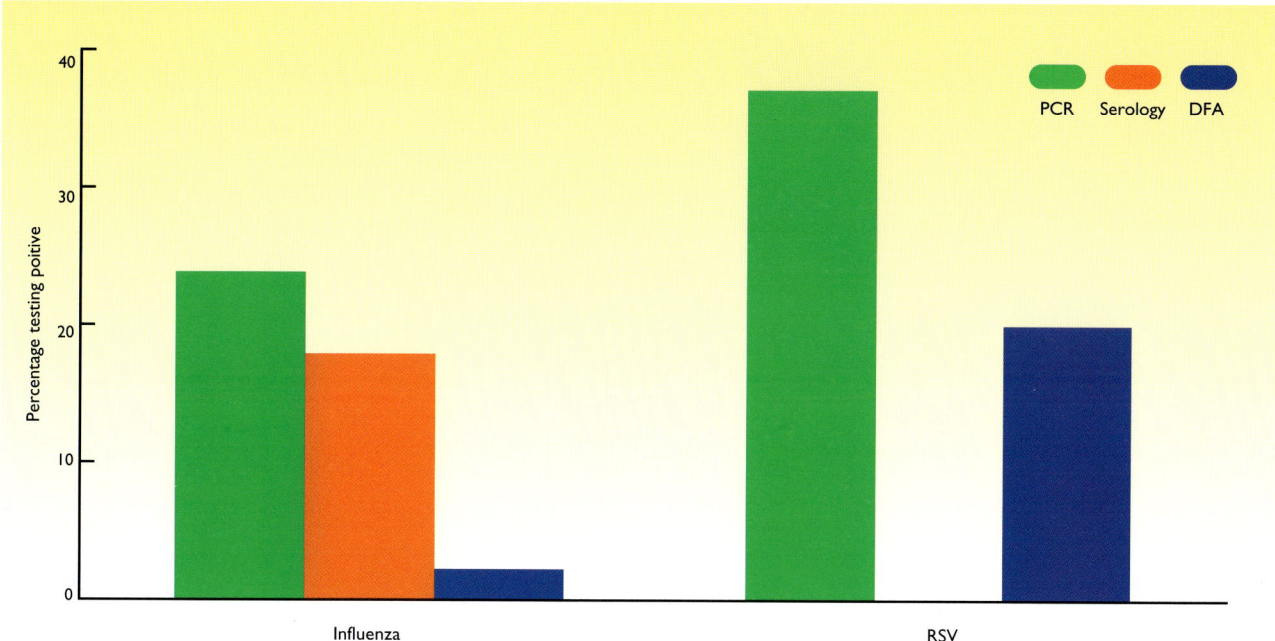

Figure 6.8 Diagnosis of infections in acute asthma. Percentage of induced sputum samples testing positive by PCR and direct fluorescence antigen (DFA) detection for influenza and respiratory syncytial virus (RSV) infections, compared with serology. Reproduced with permission from Simpson JL, Moric I, Wark PAB, *et al*. Use of induced sputum for the diagnosis of influenza and respiratory syncytial virus infections in asthma: a comparison of diagnostic techniques. *J Clin Virol* 2003;26:339–46

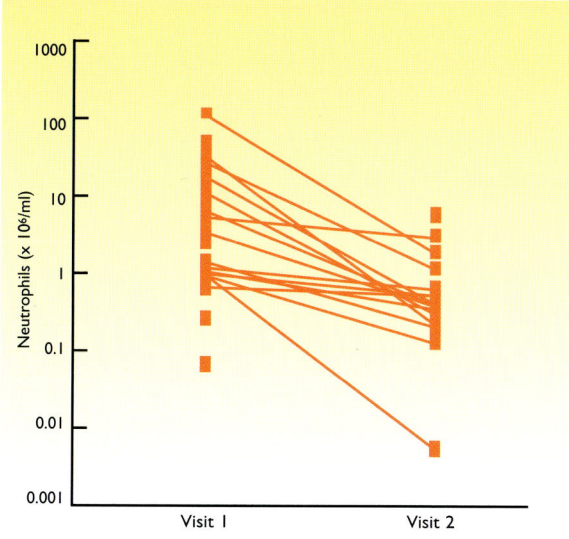

Figure 6.9 Sputum neutrophils as a proportion of non-squamous cells from children during an exacerbation (Visit 1) and when stable (Visit 2). Reproduced with permission from Norzila M, Fakes K, Henry RL, *et al*. IL-8 secretion and neutrophil recruitment accompanies induced sputum eosinophil activation in children with acute asthma. *Am J Respir Crit Care Med* 2000;161:769–74

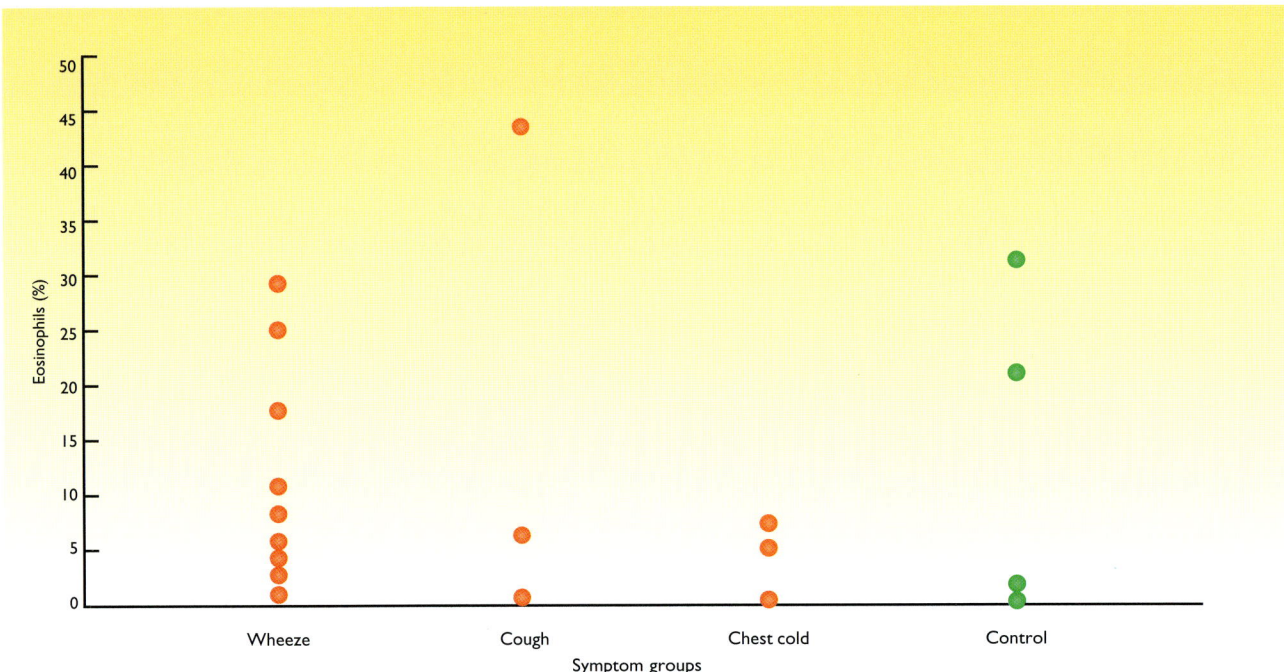

Figure 6.10 Sputum eosinophils in children with wheeze, cough, or chest colds and healthy controls. Reproduced with permission from Gibson PG, Simpson JL, Hensley MJ, *et al*. Airways eosinophilia is associated with wheeze but is uncommon in children with persistent cough and frequent chest colds. *Am J Respir Crit Care Med* 2001;164;977–81

inflammation in childhood asthma. Sputum is readily induced in children over 6 years of age. Sputum eosinophil counts are increased in children with asthma, and the degree of eosinophilia relates to clinical disease activity. Whether measurement of inflammatory markers might contribute to clinical management of childhood asthma remains to be evaluated in large studies.

CHAPTER 7

Induced sputum studies in cystic fibrosis

Janis Shute

Cystic fibrosis is the most common lethal genetically inherited disease to affect Caucasian populations. The primary autosomal recessive genetic defect is carried by 1 in 25 of the population and affects 1 in 2500 newborn babies in the UK. For babies born in the 1990s with cystic fibrosis, a life expectancy of 40 years of age is now estimated. The genetic defect lies in the expression of a protein, the cystic fibrosis transmembrane conductance regulator (CFTR), which is an anion channel expressed by epithelial cells. In the airways, the fact that the protein is either absent or defective leads to an imbalance in movement of water and electrolytes across the bronchial epithelium, leading to abnormal mucus gland secretions that are dehydrated and viscous. Such secretions invite infection, inevitably with *Pseudomonas aeruginosa*, an organism that has many immunoevasive properties and is hard to eradicate in this group of patients. A neutrophilic inflammatory response ensues. Mucus secretion from hyperplastic mucus glands is stimulated by neutrophil proteases, and self-amplifying cycles of infection and inflammation are initiated. Necrotic inflammatory cells, mainly neutrophils, release large amounts of DNA and actin, (high molecular weight polymers) in the airways that further contribute to increased mucus viscosity. The mucociliary escalator is unable to clear the resulting tenacious secretions and daily chest physiotherapy becomes a necessity.

Quantitative measures of infection and inflammation in the airways are important in the analysis of disease progression and response to therapy. Expectorated sputum can provide an accurate measure of infection and inflammation in the lower airways, but only patients with moderate to severe lung disease produce sputum spontaneously on a regular basis, limiting its use to more severe patients. Also, despite an increasing adult population, the majority of patients with cystic fibrosis are children, and many who are early in the disease process are unable to spontaneously expectorate sputum. Therefore, sputum induction is a useful diagnostic tool for the analysis of inflammation, especially in children with mild disease, and is superior to oropharyngeal culture, bronchoalveolar lavage or spontaneously expectorated sputum for the detection of infection.

THE NEED, EFFICACY AND ACCEPTABILITY OF INDUCED SPUTUM IN CYSTIC FIBROSIS

While sputum induction is usually very successful in adults, this is not necessarily the case in children. A retrospective study analyzed the proportion of children between the ages of 6 and 12 years with cystic fibrosis who were able, with the help of a respiratory therapist, to expectorate sputum spontaneously at their annual visit. The proportion

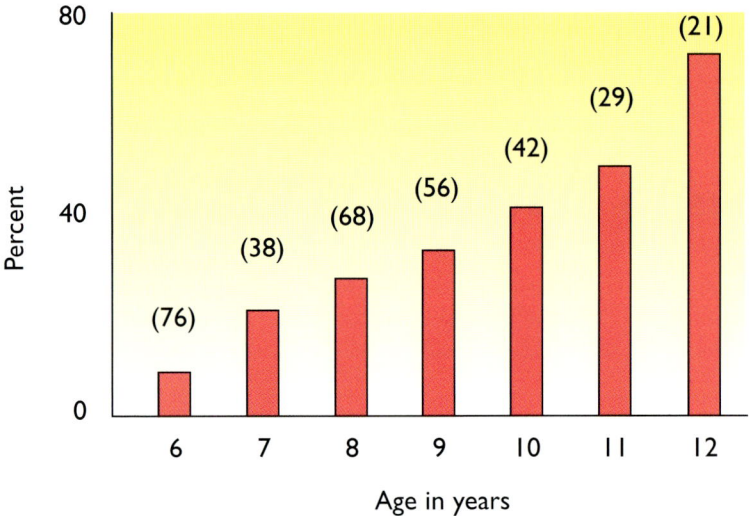

Figure 7.1 Percentage of children with cystic fibrosis who are capable of spontaneously expectorating sputum at different ages. The number of children included in the analysis is shown in parentheses. Reproduced with permission from Sagel SD, Kapsner R, Accurso FJ, *et al*. Airway inflammation in children with cystic fibrosis and healthy children assessed by sputum induction. *Am J Respir Crit Care Med* 2001;164:1425–31

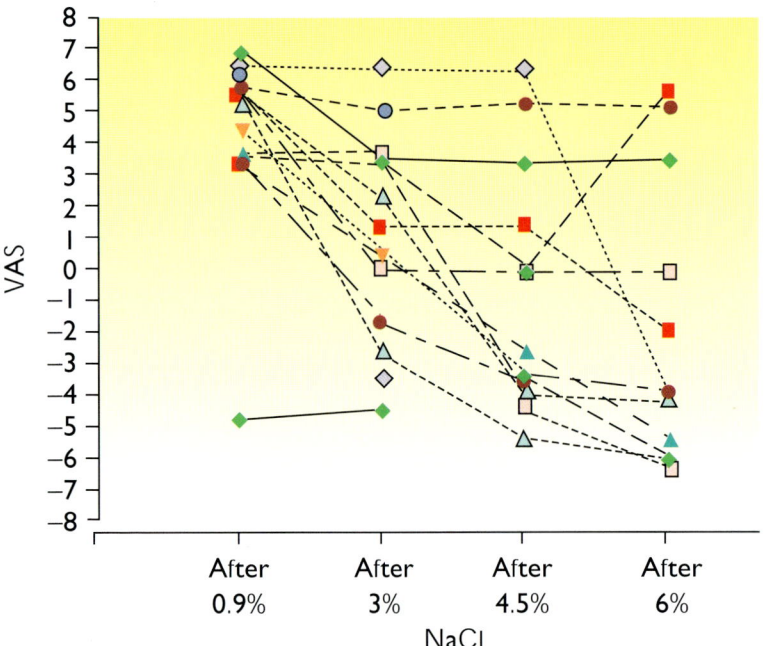

Figure 7.2 Acceptability of sputum induction expressed as visual analog score (VAS). The different symbols represent individual patients. Reproduced with permission from De Boeck K, Alifier M, Vandeputte S. Sputum induction in young cystic fibrosis patients. *Eur Resp J* 2000;16:91–4

increased with age, but was still less than 50% at 11 years of age (Figure 7.1). This observation highlights the need for an even less invasive technique to sample the airways in young children.

A systematic study in 19 children between 4 and 15 years of age, who could not spontaneously expectorate, demonstrated the possibility of inducing sputum in all patients studied and the acceptability of sputum induction with both isotonic and hypertonic saline. Increasing concentrations of sodium chloride – 0.9, 3, 4.5 and 6% – were nebulized and 20 ml of each concentration inhaled over

a 5-min period. Between inhalations children were encouraged to cough up sputum, although it was noticed that the younger children were sometimes swallowing the secretions and special coaching was needed. Sputum induction with hypertonic saline was more successful than with normal saline. One child produced sputum after inhaling 0.9% NaCl, six children after 3% NaCl, one after 4.5% NaCl, and 11 after 6% NaCl, but hypertonic saline was also considered more unpleasant (Figure 7.2). Hypertonic salt solutions were scored as significantly more unpleasant than isotonic saline, and

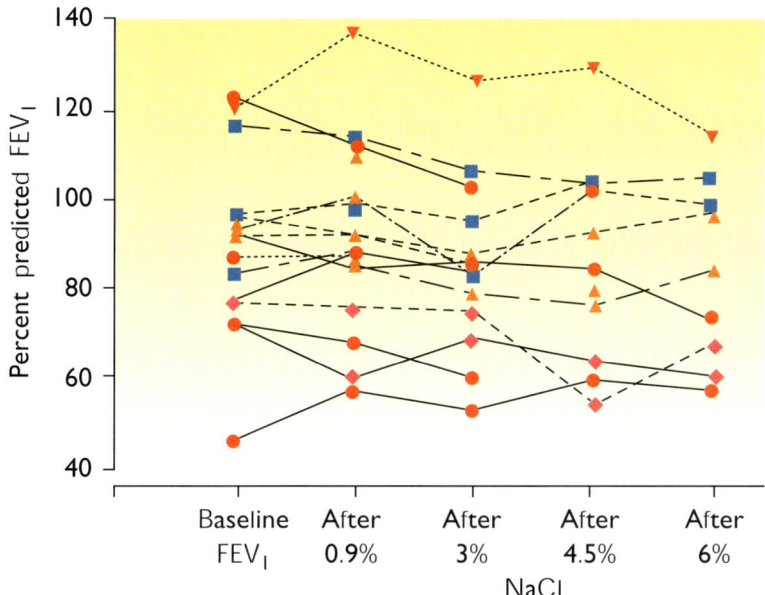

Figure 7.3 Changes in forced expiratory volume in 1 sec (FEV$_1$) during sputum induction. The different symbols represent individual patients. Reproduced with permission from De Boeck K, Alifier M, Vandeputte S. Sputum induction in young cystic fibrosis patients. *Eur Resp J* 2000;16:91–4

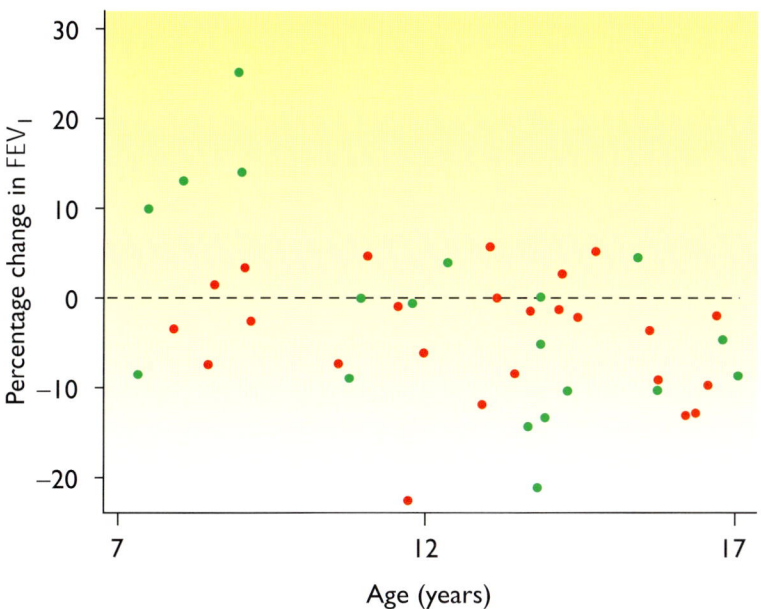

Figure 7.4 Correlation between increase in patient's age and percentage drop in forced expiratory volume in 1 sec (FEV$_1$) with 7% hypertonic saline; $p = 0.006$; $R = -0.41$. Reproduced with permission from Suri R, Marshall LJ, Shute JK, *et al.* Effects of rhDNase and hypertonic saline on airway inflammation in children with cystic fibrosis. *Am J Respir Crit Care Med* 2002;166: 352–5

children complained about the salty taste starting at 3% NaCl solutions.

In a study of 45 children (7–17 years of age) who inhaled 5 ml of 7% hypertonic saline for 12 min, only five patients complained of the salty taste, although this was not severe enough for them to stop the test. However, hypertonic saline made all patients cough during administration. It was concluded that sputum induction is successful, safe, and acceptable in cystic fibrosis patients who do not expectorate spontaneously, and can be performed from 4 years of age.

SAFETY OF INDUCED SPUTUM IN CHILDREN AND ADULTS WITH CYSTIC FIBROSIS

Most studies have indicated that sputum induction is safe in both children and adults with cystic

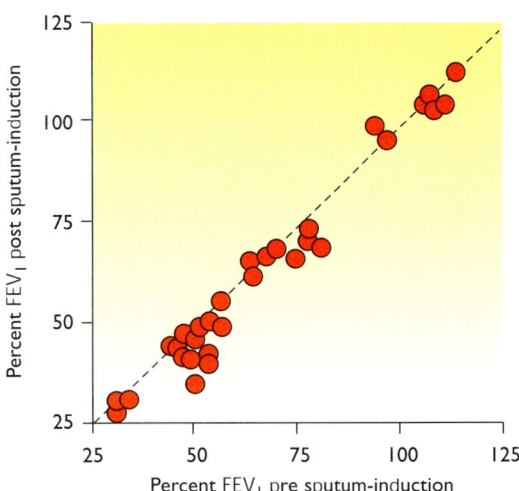

Figure 7.5 Percent predicted FEV₁ before and after sputum induction in ten adult patients with cystic fibrosis during four visits at 3-week intervals (39 procedures). Reproduced with permission from Ordonez CL, Stulbarg M, Boushey HA, *et al*. Effect of clarithromycin on airway obstruction and inflammatory markers in induced sputum in cystic fibrosis: a pilot study. *Pediatr Pulmonol* 2001;32: 29–37

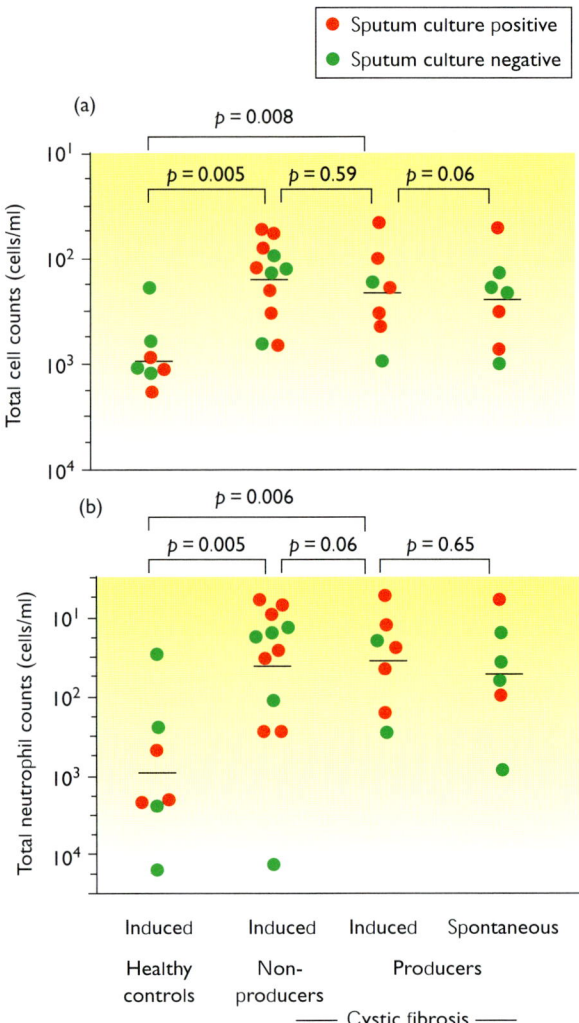

Figure 7.6 Total cell counts (a) and neutrophil counts (b) in sputum from healthy control subjects and children with cystic fibrosis. Two groups of patients were studied: sputum producers (those capable of spontaneously expectorating sputum) and non-producers. Bars indicate mean values for each group. The total cell counts and neutrophil counts were significantly higher in the patient groups (significance values as shown). Reproduced with permission from Sagel SD, Kapsner R, Accurso FJ, *et al*. Airway inflammation in children with cystic fibrosis and healthy children assessed by sputum induction. *Am J Respir Crit Care Med* 2001;164:1425–31

fibrosis, although a small but significant decrease in forced expiratory volume in 1 second (FEV_1) following sputum induction is often recorded. It has been suggested that in some patients the decrease in FEV_1 might be due to mucus plugging rather than bronchoconstriction. For this reason, in addition to premedicating patients with a short-acting bronchodilator, it is recommended that pulmonary function (FEV_1 or peak flow) be monitored throughout the procedure.

It is important to note that premedication with bronchodilators does not prevent bronchoconstriction. Thus, in one study of 19 children who inhaled a nebulized solution of increasing concentrations of sodium chloride until sputum was produced, bronchodilator pretreatment with salbutamol did not prevent a small but significant drop in FEV_1 (Figure 7.3). The mean change (range) in post-salbutamol FEV_1 (precent predicted) after the test was -7% (range -24–16%; $p < 0.03$). In two children the drop exceeded 20%, but in six children the FEV_1 was improved at the final saline concentration.

In a study of 45 children who inhaled 5 ml of 7% hypertonic saline, 30 patients had a fall in FEV_1 and two experienced a clinically significant drop in $FEV_1 > 15\%$. Furthermore, it was observed that there was a positive correlation between

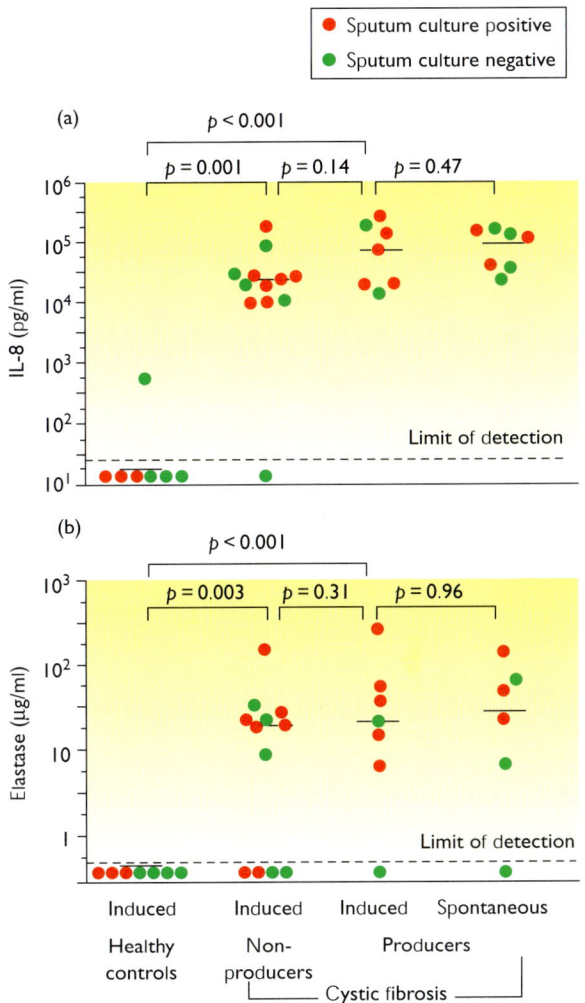

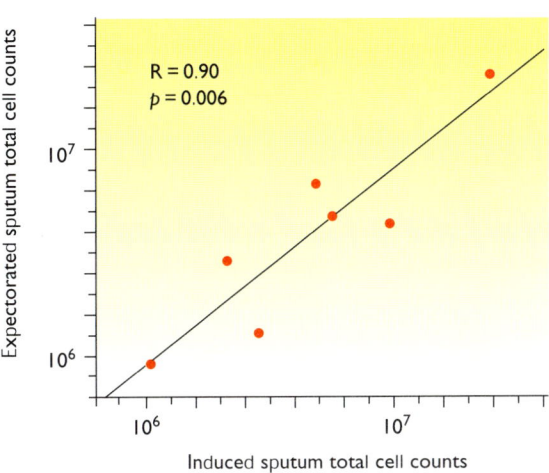

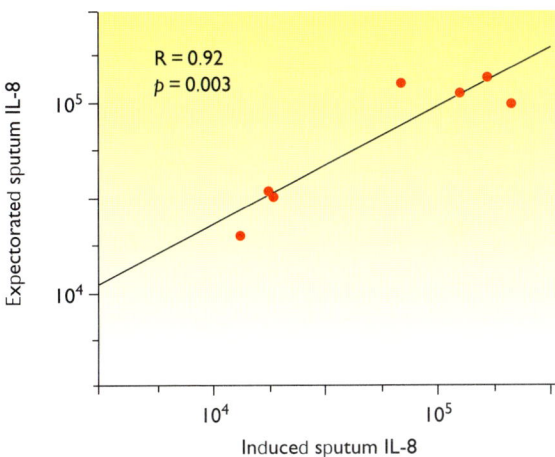

Figure 7.7 Concentrations of IL-8 (a) and neutrophil elastase activity (b) in sputum from healthy subjects and children with cystic fibrosis. Two groups of patients were studied: sputum producers and non-producers. Bars indicate median values for each group and dotted lines indicate the limit of detection of the assays. IL-8 and elastase activity were significantly higher in both cystic fibrosis groups (significance values as shown). Reproduced with permission from Sagel SD, Kapsner R, Accurso FJ, *et al.* Airway inflammation in children with cystic fibrosis and healthy children assessed by sputum induction. *Am J Respir Crit Care Med* 2001;164:1425–31

Figure 7.8 Correlation between total cell counts (top) and IL-8 measurements in induced sputum and matched expectorated sputum specimens from children with cystic fibrosis. Reproduced with permission from Sagel SD, Kapsner R, Accurso FJ, *et al.* Airway inflammation in children with cystic fibrosis and healthy children assessed by sputum induction. *Am J Respir Crit Care Med* 2001;164: 1425–31

increase in age and the percentage drop in FEV_1 over the age range 7–17 years (Figure 7.4). Older children generally experienced a greater percentage drop in FEV_1 despite the use of short-acting bronchodilators. There was no correlation between the percentage change in FEV_1 with hypertonic saline and disease severity, assessed as percent of predicted FEV_1 at baseline.

The safety of inducing sputum in adult patients with cystic fibrosis has been confirmed. In ten adult patients aged 19–26 years, with moderate to severe airflow obstruction, sputum was induced

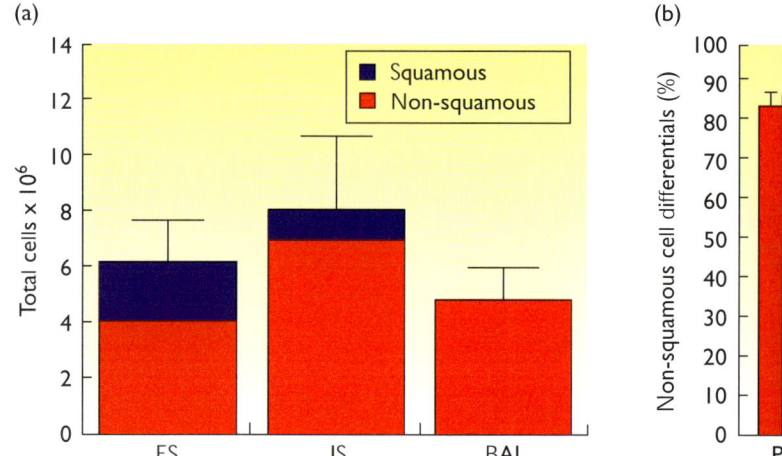
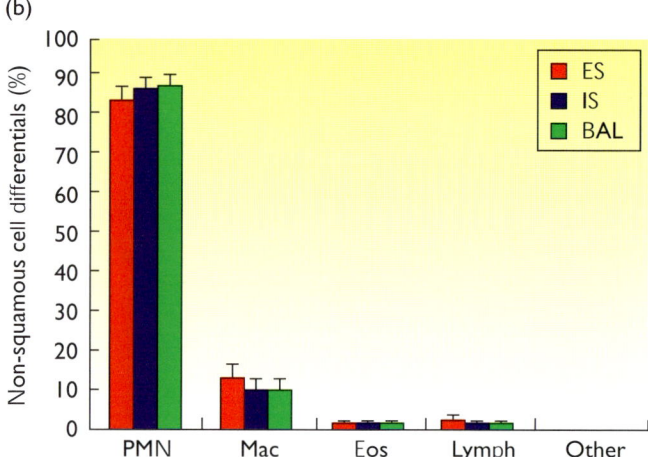

Figure 7.9 Comparison between sputum and bronchoalveolar lavage (BAL). (a) Total cell counts (expressed per ml of original sample). There was no significant difference in the number of cells collected. The percentage of non-squamous, non-epithelial cells was different between spontaneously expectorated sputum (ES) and both induced sputum (IS) and BAL ($p = 0.03$ and $p = 0.006$, respectively); (b) differential cell counts of non-squamous cells. The cell populations were similar with each of the three sampling procedures. PMN, polymorphonuclear cells; Mac, macrophages; Eos, eosinophils; Lymph, lymphocytes. Reproduced with permission from Henig NR, Tonelli MR, Aitken ML, *et al*. Sputum induction as a research tool for sampling the airways of subjects with cystic fibrosis. *Thorax* 2001;56:306–11

with 3% hypertonic saline inhaled for 12 min. The mean percent predicted FEV_1 before sputum induction was 65% (range 31–113%); after sputum induction it was 61% (range 28–113%) (Figure 7.5). In no patient was the procedure terminated before 12 min of inhalation, and the greatest decline in FEV_1 over 39 procedures was 15%, a fall that was reversed by salbutamol.

INFLAMMATORY CELLS AND MEDIATORS IN INDUCED VERSUS EXPECTORATED SPUTUM

Inflammatory cell counts and biochemical indices of inflammation in induced sputum are similar to those previously reported for spontaneously expectorated sputum in children with cystic fibrosis. In a study directly comparing cellular and biochemical indices of inflammation, there was no significant difference in these measurements in induced sputum compared with spontaneously expectorated sputum in children. Even in children who do not normally produce sputum, total cell counts, as well as neutrophil counts, in induced sputum are significantly increased compared with

healthy children (Figure 7.6), and the degree of inflammation is not significantly different from children with cystic fibrosis who have a productive cough and can, therefore, produce sputum spontaneously.

Similarly, interleukin-8 (IL-8) levels and neutrophil elastase activity are significantly higher in the patient group than in healthy children, but are not different between patients with cystic fibrosis who are capable of producing sputum spontaneously and those who are not (Figure 7.7).

In sputum producers, i.e. those who can expectorate spontaneously, there is a highly significant correlation between total cell counts and IL-8 levels in matched induced and expectorated samples (Figure 7.8); this indicates that analysis of induced sputum may provide useful outcome measures in clinical trials in children with cystic fibrosis. Additionally, a significant difference in the amount of induced sputum (8.3 ± 2.9 g) compared with the mean weight of sputum expectorated spontaneously (0.5 ± 0.1 g) allows for measurement of a more extensive range of inflammatory markers. Sputum induction in adult patients with cystic fibrosis similarly results in

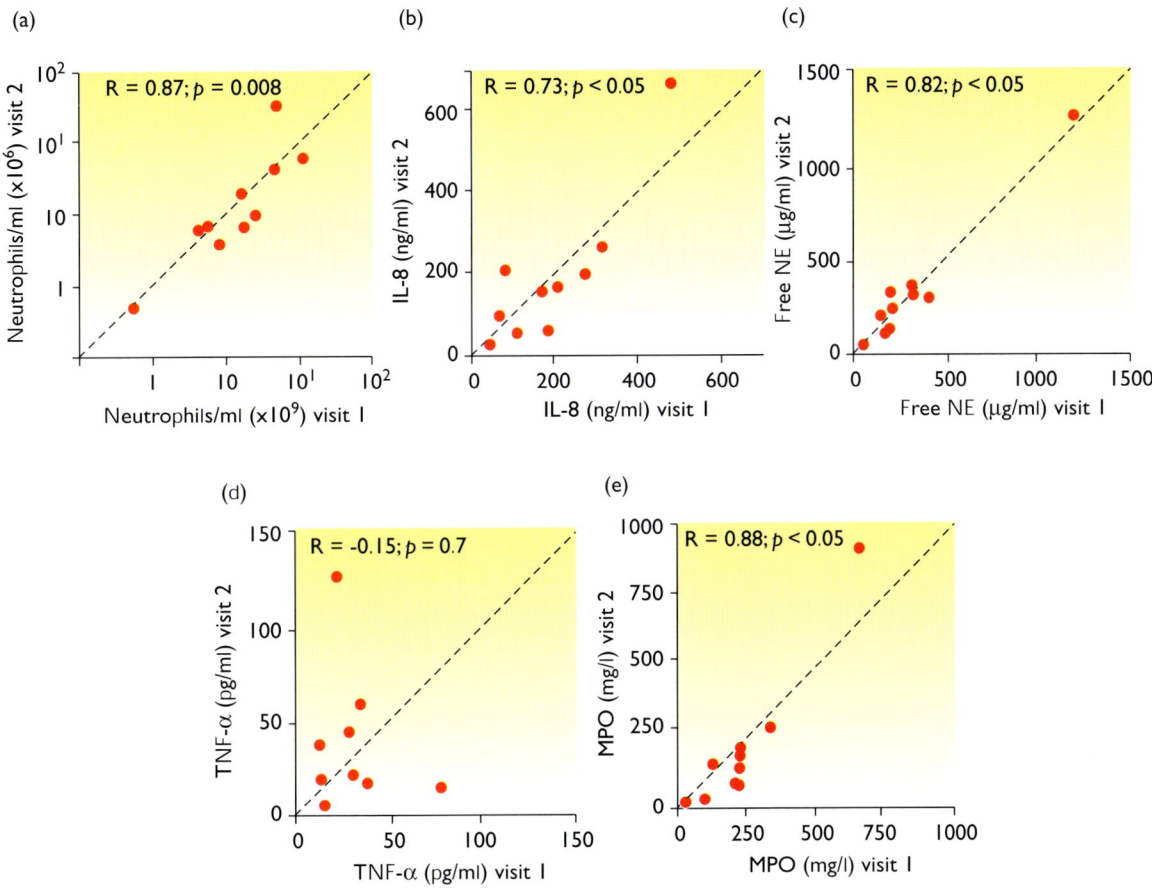

Figure 7.10 Reproducibility of (a) neutrophil number; (b) IL-8 levels; (c) neutrophil elastase (NE) activity; (d) TNF-α levels; and (e) myeloperoxidase (MPO) levels in induced sputum obtained on two visits with a 3-week interval. Dashed line represents the line of identity; R, intraclass correlation coefficient. Reproduced with permission from Ordonez CL, Stulbarg M, Boushey HA, *et al.* Effect of clarithromycin on airway obstruction and inflammatory markers in induced sputum in cystic fibrosis: a pilot study. *Pediatr Pulmonol* 2001;32:29–37

significantly larger sample volumes than spontaneously expectorated samples.

COMPARISONS BETWEEN INDUCED SPUTUM AND BRONCHOALVEOLAR LAVAGE AS RESEARCH TOOLS

Fiberoptic bronchoscopy is an invasive technique that is believed to sample the alveolar space, rather than the airways alone. Also it is recognized that the procedure can itself induce changes in inflammatory mediator levels in the airways. Thus, in addition to being time consuming and costly, it is not well suited to serial sampling of the airways in longitudinal studies. As a research tool, the inexpensive and non-invasive technique of sputum induction therefore appears to have a number of advantages over bronchoalveolar lavage (BAL).

Comparative studies have shown that sputum induction with 3% hypertonic saline for 12 min is well tolerated and preferred to BAL by adult patients with cystic fibrosis. Contamination of samples with squamous epithelial cells reduces the validity of total and differential cell counts in the samples. However, the number of non-squamous, non-epithelial cells is not different when comparing induced sputum and BAL, and both sputum

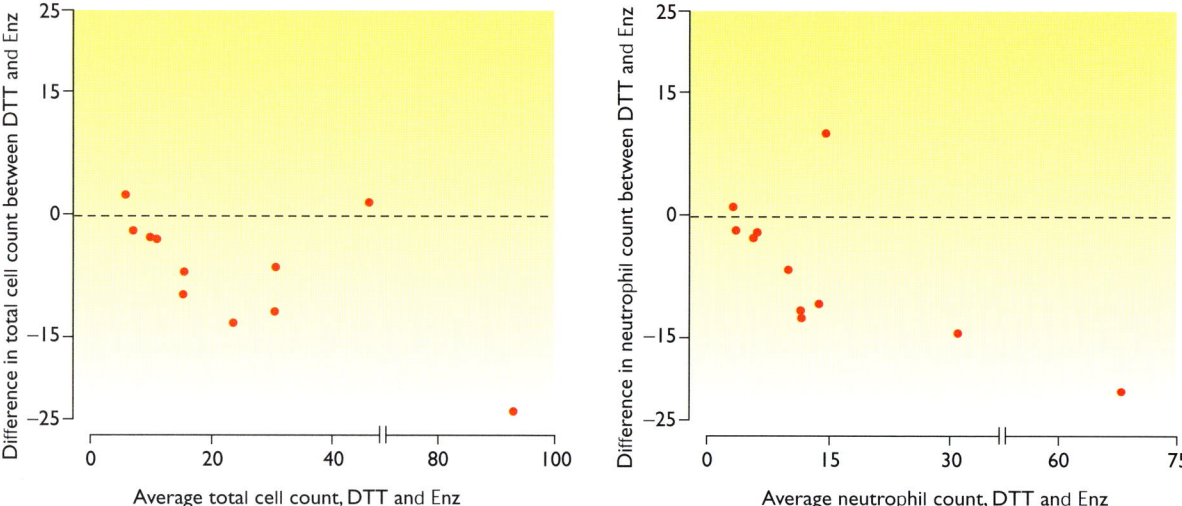

Figure 7.11 Agreement between total cell count and neutrophil counts in cystic fibrosis sputum processed using dithiothreitol (DTT) or an enzyme mixture, Enz (DNase I, hyaluronidase, galactosidase) to disperse cells. The average cell count is plotted against the difference in cell counts using each technique. Data points are consistently below the zero point (broken line) indicating poor agreement between the two methods, with DTT counts consistently lower than Enz counts ($p < 0.05$). Reproduced with permission from Cai Y, Carty K, Gibson P, Henry R. Comparison of sputum processing techniques in cystic fibrosis. *Pediatr Pulmonol* 1996;22:402–7

induction and BAL generate samples with significantly less squamous cell contamination than is seen in sputum expectorated spontaneously (Figure 7.9). Differential cell counts (excluding squamous cells) are similar for all methods (Figure 7.9).

The disadvantages of sputum induction compared with BAL are that sequential sampling of the same subsegmental airway is not possible, and the technique cannot be taught to very young children less than 4 years of age.

REPRODUCIBILITY

Reproducibility of inflammatory markers in induced sputum in cystic fibrosis patients has been generally good, as shown in studies comparing samples collected from clinically stable adult patients on two separate visits, 3 weeks apart. The reproducibility of measurements has been high for neutrophil numbers, levels of IL-8 and MPO, and neutrophil elastase activity, but less for tumor necrosis factor (TNF)-α (Figure 7.10).

INFECTION

Comparisons of sputum induction with sputum expectoration and BAL for the recovery of bacterial and fungal pathogens in adult patients have indicated that sputum induction detects pathogens not detected using the other techniques. Overall, the sensitivity of isolation of organisms from induced sputum is slightly, but not statistically, better than expectorated sputum or BAL. In addition, higher colony counts are recovered in induced sputum. In young cystic fibrosis patients, a greater density of colony counts can also be recorded from induced sputum compared with nasopharyngeal aspirates. It has been suggested that sputum induction may be useful as both a research tool and a clinical method for identifying respiratory pathogens in cystic fibrosis.

PROCESSING

Whether induced or spontaneously expectorated, sputum from patients with cystic fibrosis is unique

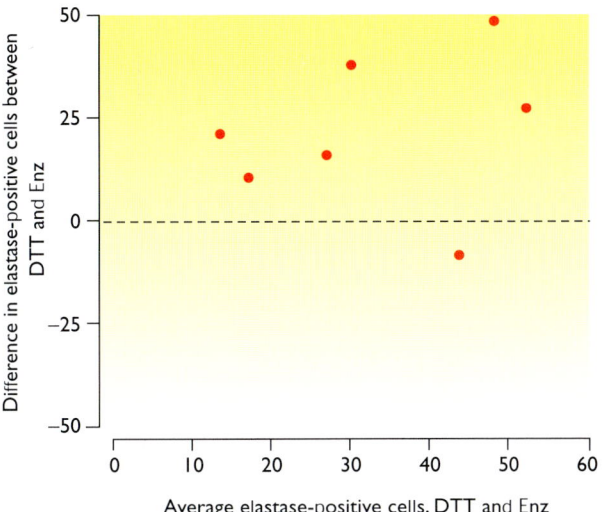

Figure 7.12 Agreement between elastase-positive neutrophils (%) in cystic fibrosis sputum processed using dithiothreitol (DTT) or an enzyme mixture (Enz) to disperse cells. Data points are consistently above the zero point (broken line), indicating poor agreement between the two methods, with DTT counts consistently higher than Enz counts ($p < 0.05$). Reproduced with permission from Cai Y, Carty K, Gibson P, Henry R. Comparison of sputum processing techniques in cystic fibrosis. *Pediatr Pulmonol* 1996;22:402–7

in the high levels of DNA (1.6–28 mg/ml), mucin (10–50 mg/ml) and actin (0.1–5 mg/ml) that contribute to its highly viscous nature. DNA and mucin concentrations are higher in cystic fibrosis than in asthma, which indicates that the processing techniques described in previous chapters may not be suitable for samples from patients with cystic fibrosis. For example, it has been shown that following processing of sputum from patients with cystic fibrosis using a mixture of DNase, β-galactosidase and hyaluronidase enzymes (all at 10 U/ml), total cell counts and neutrophil counts are significantly higher than the counts obtained using dithiothreitol (DTT) (Figure 7.11).

However, despite good recovery of cells, such enzyme treatment has been recognized to damage some antigens. The immunoreactivity of cellular neutrophil elastase is significantly decreased by galactosidase treatment (Figure 7.12) and it has been speculated that the same may be true of the soluble mediator. Conversely, DNase treatment of cystic fibrosis sputum *in vitro* increases detectable levels of IL-8 and MPO. The limitation of these methods may relate to specific antigen reactivity, but the observations emphasize the importance of testing the effects of cell dispersal reagents on the subsequent detection of sputum components before using these measures in clinical trials.

When markers of inflammation are the outcome measure of a clinical trial, the procedure of rapidly freezing whole sputum at -80°C on the study day can be adopted, thawing samples in batches at the end of the study, and processing with an equal volume of phosphate buffered saline (PBS). This procedure generates a soluble phase that has not been subjected to the potential effects of DTT on mucin-bound mediators. Processing sputum with DTT can significantly increase the detectable levels of neutrophil elastase activity and myeloperoxidase, ECP and IL-8 immunoreactivity compared with levels detected using PBS. In addition, processing with PBS eliminates potential effects of DTT in subsequent immunoassays.

In summary, analysis of induced sputum is a useful and non-invasive method for studying airway inflammation in children and adults with cystic fibrosis. Methods of processing sputum with DTT may not give optimal results for cell counts or mediator measurements when applied to cystic fibrosis sputum. Investigators should consider that samples may need to be split and portions processed differently, depending on the marker or mediator of interest. The optimum method for processing sputum from cystic fibrosis patients for immunocytochemistry remains to be determined.

CHAPTER 8

The use of induced sputum in clinical trials

Paula Rytilä

The past two decades have seen a proliferation of clinical trials whose primary objective has been to provide evidence of effectiveness and help place drugs within national and international treatment guidelines. Increasingly, evidence is being sought to provide proof of the anti-inflammatory action of drugs and improve understanding of how drugs work. Asthma and chronic obstructive pulmonary disease (COPD) are both chronic inflammatory diseases of the airways in which inflammation is a prominent feature and is broadly associated with clinical disease. In both conditions, it is difficult to assess the nature or degree of airway inflammation from clinical parameters alone. Direct indices of airway inflammation can correlate with clinical parameters such as symptoms, degree and variability of airflow limitation, and airway responsiveness that are normally monitored during a clinical trial. However, the correlation is not good for all parameters, and changes in various inflammatory and clinical parameters of disease activity may not occur simultaneously.

Bronchoscopy has been used to obtain bronchial biopsy and bronchioalveolar lavage (BAL) samples in clinical trials. However, the invasiveness of these procedures, the associated costs, and significant restriction of their use to patients with mild or moderately severe disease has meant that bronchoscopy is not well suited to monitoring airway

inflammation in large clinical trials. As discussed in Chapter 1, the methods for sputum induction and sample processing are reasonably safe and reproducible. In addition, sputum cell counts, especially the number or percentage of sputum eosinophils, have been well validated in terms of responsiveness to intervention. For these reasons induced sputum analysis is increasingly used in clinical trials.

A number of methodological issues discussed earlier in this atlas need to be emphasized when applying induced sputum in the context of clinical trials. The length of sputum induction might change the numbers of inflammatory cells and the interval between consecutive inductions also needs to be taken into account. These parameters need to be standardized in all clinical trials. Also, any delay in sputum processing might influence the results and should be minimized. Attempts have been made recently to simplify further both the induction and processing of samples for use in epidemiological studies; in the future these may also become useful for clinical trials.

Several clinical trials in asthmatic patients and recently in COPD patients have assessed the effects of inhaled and oral corticosteroids on parameters in induced sputum (Figures 8.1–8.4). A number of studies have combined sputum induction with other techniques used to measure airway inflammation indirectly, such as measurement of

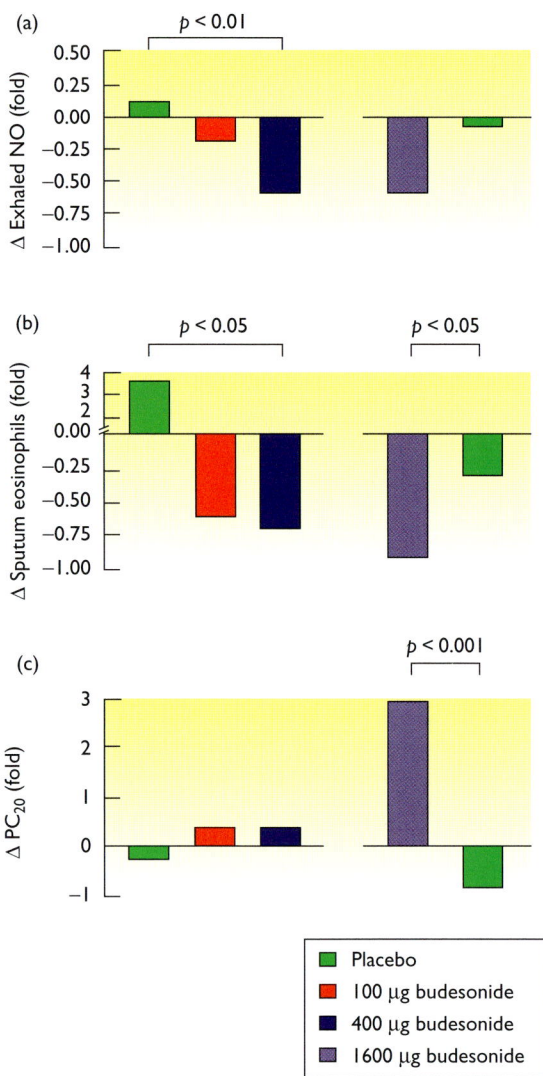

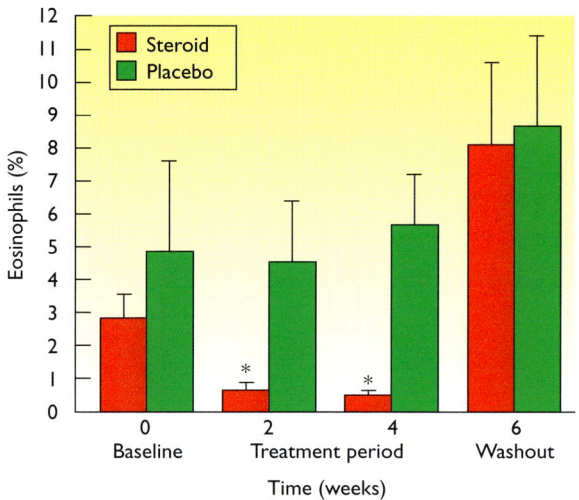

Figure 8.2 The effect of the inhaled corticosteroid fluticasone in mild asthma. Eosinophil counts (shown as percentages of total inflammatory cells) in induced sputum at baseline, at weeks 2 and 4 of treatment, and after 2 weeks of washout in the steroid-treated (fluticasone 500 μg twice daily) and placebo-treated patients. *Significant difference between the two groups. Reproduced with permission from van Rensen ELJ, Straathof KCM, Sterk PJ, et al. Effect of inhaled steroids on airway hyperresponsiveness, sputum eosinophils, and exhaled nitric oxide levels in patients with asthma. *Thorax* 1999;54:403–8

Figure 8.1 The effects of increasing doses of the inhaled corticosteroid, budesonide, on airway inflammation and responsiveness. Changes in (a) exhaled nitric oxide (NO); (b) sputum eosinophil numbers; and (c) methacholine airway responsiveness (PC_{20}) in response to 4 weeks' treatment with low-dose (100 or 400 μg) budesonide are shown on the left and high-dose (1600 μg) budesonide on the right. Bars represent mean changes from baseline for NO, median change from baseline for eosinophils and changes in geometric means for PC_{20}. Reproduced with permission from Jatakanon A, Kharitonov S, Lim S, Barnes PJ. Effect of differing doses of inhaled budesonide on markers of airway inflammation in patients with mild asthma. *Thorax* 1999;54:108–14

nitric oxide (NO) in exhaled air and airway responsiveness (Figures 8.1, 8.3 and 8.4).

A study of 25 patients with mild asthma treated for 4 weeks with 500 μg of inhaled fluticasone

propionate or placebo compared within-subject changes in sputum eosinophils with those in exhaled NO and airway responsiveness. Although the treatment led to an improvement in all the parameters studied (Figure 8.2), the study failed to find a correlation between steroid-induced changes in these various parameters (Figure 8.3 and 8.4). In another study of 14 patients with mild asthma, changes in eosinophils in induced sputum, BAL and bronchial biopsies, and exhaled NO, after treatment with either 800 μg of budesonide or placebo were assessed in a double-blind randomized cross-over trial of 12 weeks (Figures 8.5 and 8.6). Again, although treatment caused significant changes in exhaled NO and eosinophilia in both sputum and bronchial biopsies, there was no correlation between the different measurements of airway inflammation. This suggests that various inflammatory measurements can provide complementary information during a clinical trial involving an anti-inflammatory drug.

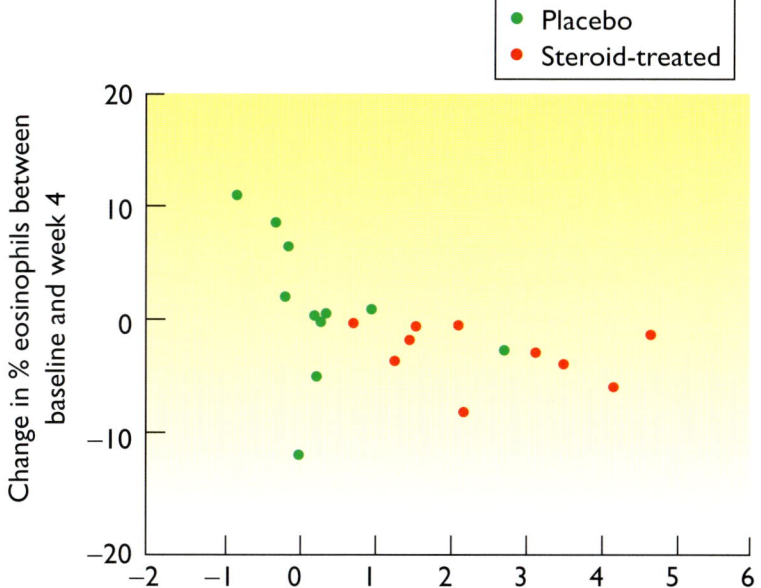

Figure 8.3 There is a lack of relationship (R < 0.56, p > 0.15) between changes in sputum eosinophil counts and changes in PC_{20} histamine at week 4 (both compared with baseline) in a study involving 25 patients with mild asthma. Steroid treatment was fluticasone 500 μg twice daily. Reproduced with permission from van Rensen ELJ, Straathof KCM, Sterk PJ, *et al*. Effect of inhaled steroids on airway hyperresponsiveness, sputum eosinophils, and exhaled nitric oxide levels in patients with asthma. *Thorax* 1999;54:403–8

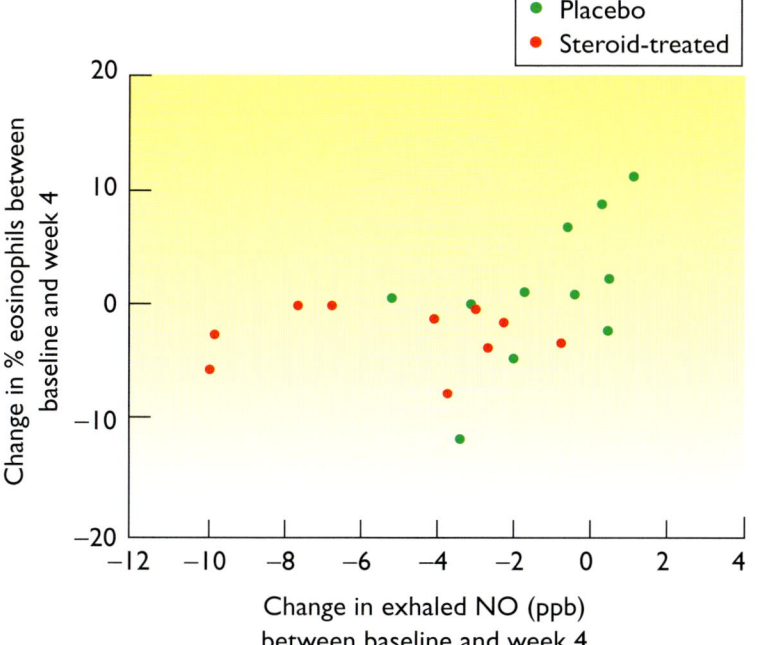

Figure 8.4 There is a lack of relationship (R < 0.56, p > 0.15) between changes in sputum eosinophils and changes in exhaled nitric oxide (NO) levels at week 4 (both compared with baseline) in a study involving 25 patients with mild asthma. Steroid treatment was fluticasone 500 μg twice daily. Reproduced with permission from van Rensen ELJ, Straathof KCM, Sterk PJ, *et al*. Effect of inhaled steroids on airway hyperresponsiveness, sputum eosinophils, and exhaled nitric oxide levels in patients with asthma. *Thorax* 1999;54:403–8

There have relatively few studies applying induced sputum to study the effects of treatment in COPD. One such example is a study involving 13 COPD patients treated with 500 μg of fluticasone propionate or placebo in a double-blind crossover study of 4-week treatment. Inflammatory cells, cytokines and proteases were also analyzed (Figures 8.7 and 8.8). This study showed no effect of a moderate dose of inhaled corticosteroid on any of the studied parameters, a finding that has been

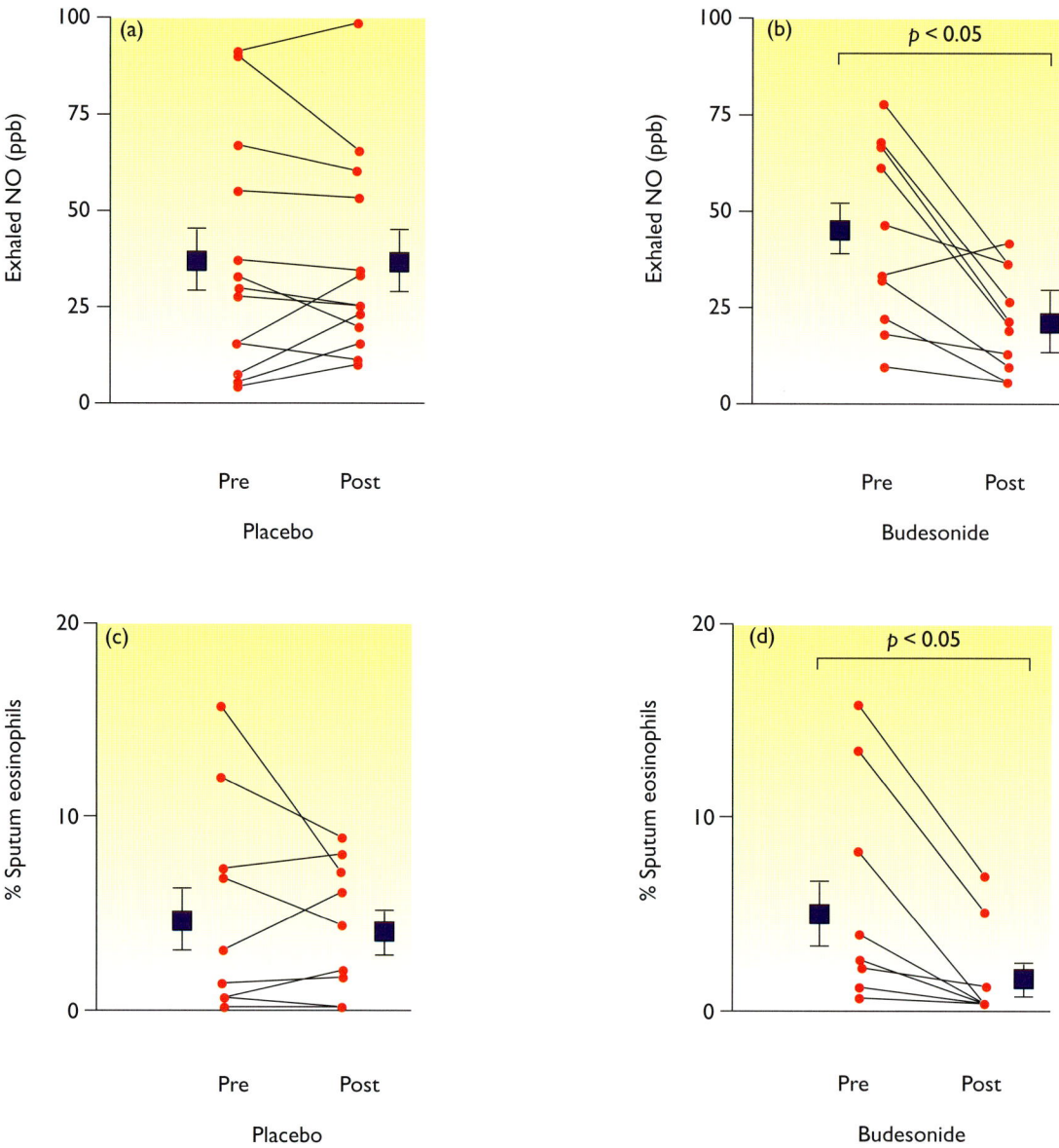

Figure 8.5 The effect of inhaled budesonide (800 μg twice daily) on exhaled nitirc oxide (NO) (a, b) and sputum eosinophils in 14 mild asthma patients in a randomized cross-over study. Each treatment was administered for 4 weeks, separated by a 4-week washout period. Results are shown as individual data pre-and post-treatment with placebo and budesonide, and means ± SEM are also shown. Reproduced with permission from Lim S, Jatakanon A, Barnes PJ, *et al.* Effect of inhaled budesonide on lung function and airway inflammation. Assessment by various inflammatory markers in mild asthma. *Am J Respir Crit Care Med* 1999;159:22–30

used to suggest that these drugs are not effective for COPD treatment. Whether this applies to other markers of inflammation and mucin generation remains to be established. It is also unclear whether a longer period of treatment and/or higher doses of corticosteroid are needed for effects to be seen.

Induced sputum in increasingly used also in multicenter trials, be it in the context of clinical trials or basic studies of disease pathogenesis. For proper conduct of such studies and their correct interpretation, it is essential that all centers participating in a multicenter trial apply a well

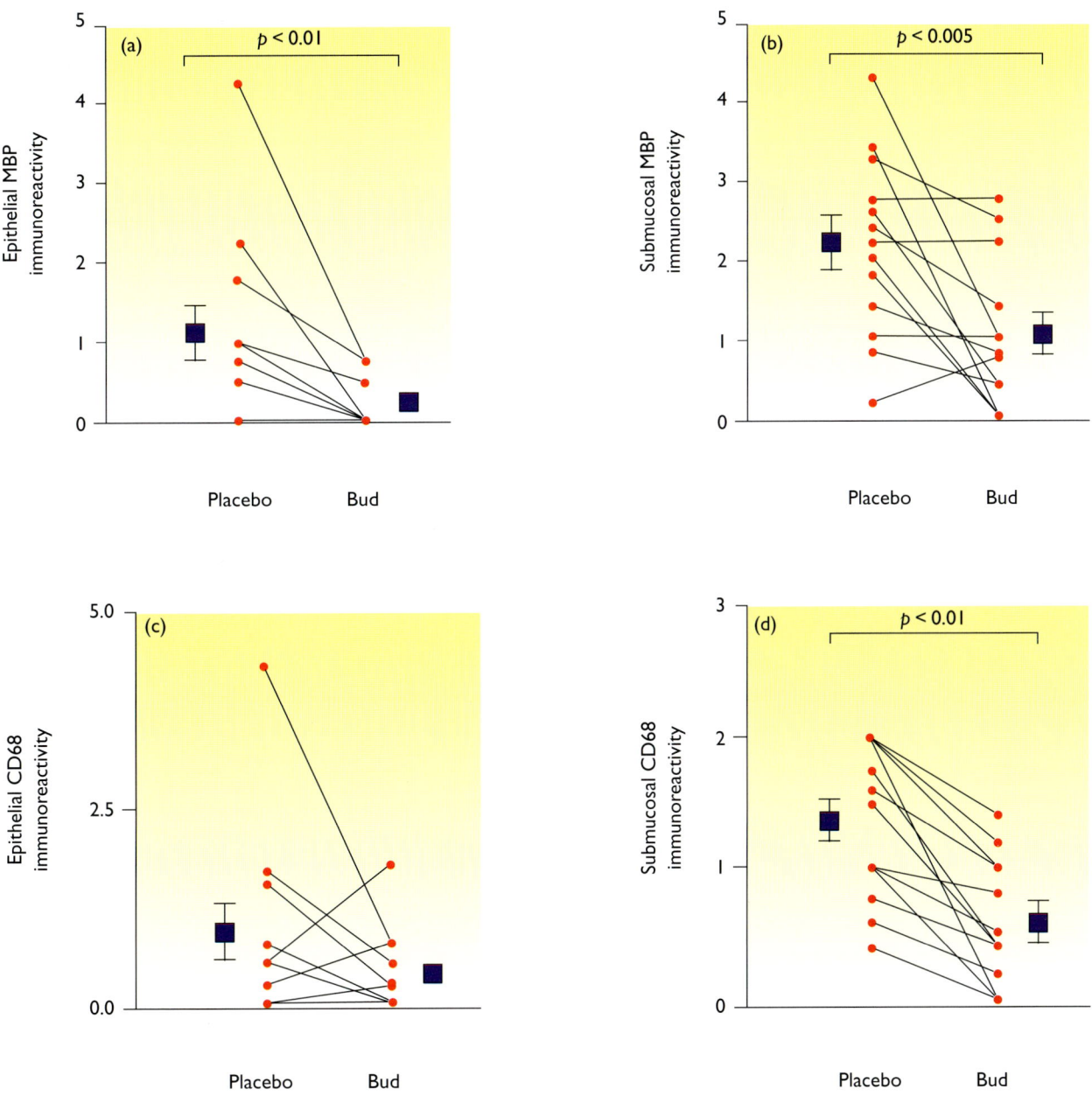

Figure 8.6 The effect of inhaled budesonide (800 μg twice daily) on major basic protein (MBP)-positive eosinophils (a, b) and CD68-positive macrophages (c, d) in the epithelium and submucosa in bronchial biopsies from 14 mild asthma patients in a randomized cross-over study. Each treatment was administered for 4 weeks, separated by a 4-week washout period. The results are shown as individual data pre-and post-treatment with placebo and budesonide, and mean ± SEM are also shown. Reproduced with permission from Lim S, Jatakanon A, Barnes PJ, *et al*. Effect of inhaled budesonide on lung function and airway inflammation. Assessment by various inflammatory markers in mild asthma. *Am J Respir Crit Care Med* 1999;159:22–30

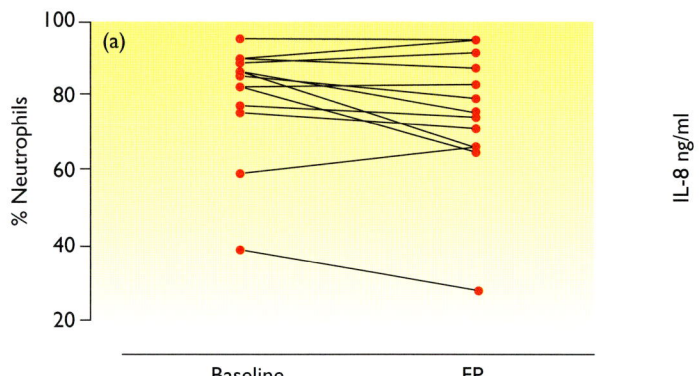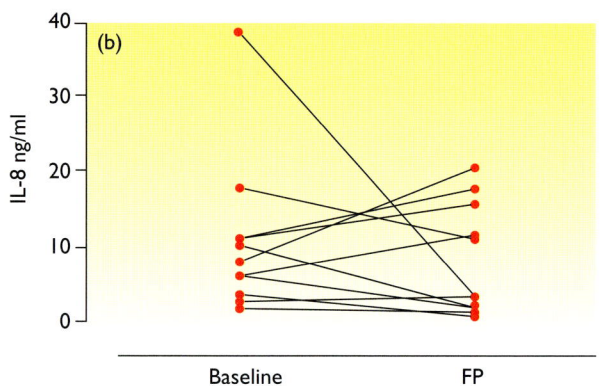

Figure 8.7 A study of the effects of 4 weeks of treatment with fluticasone propionate (FP) 500 µg twice daily on inflammatory indices in induced sputum of patients with COPD. The treatment had no effect on sputum neutrophil counts (a) or sputum IL-8 levels (b). Reproduced with permission from Culpitt SV, Maziak W, Barnes PJ, *et al.* Effect of high dose inhaled steroid on cells, cytokines, and proteases in induced sputum in chronic obstructive pulmonary disease. *Am J Respir Crit Care Med* 1999;160:1635–9

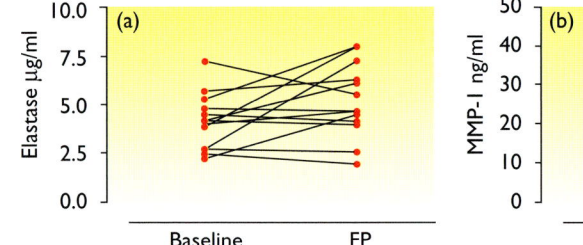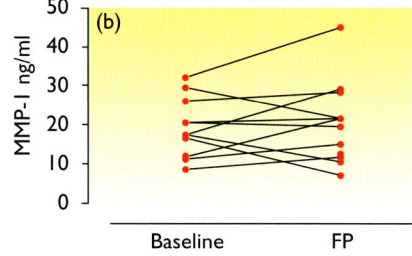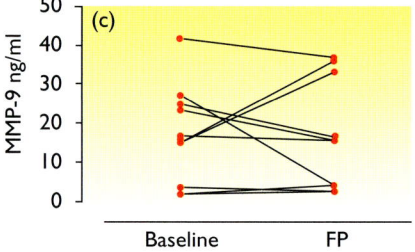

Figure 8.8 A study of the effects of 4 weeks of treatment with fluticasone propionate (FP), 500 µg twice daily, on proteases in induced sputum of patients with COPD. The treatment had no effect on total elastase activity (a), matrix metalloproteinase (MMP)-1 (b), and MMP-9 (Panel C). Reproduced with permission from Culpitt SV, Maziak W, Barnes PJ, *et al.* Effect of high dose inhaled steroid on cells, cytokines, and proteases in induced sputum in chronic obstructive pulmonary disease. *Am J Respir Crit Care Med* 1999;160:1635–9

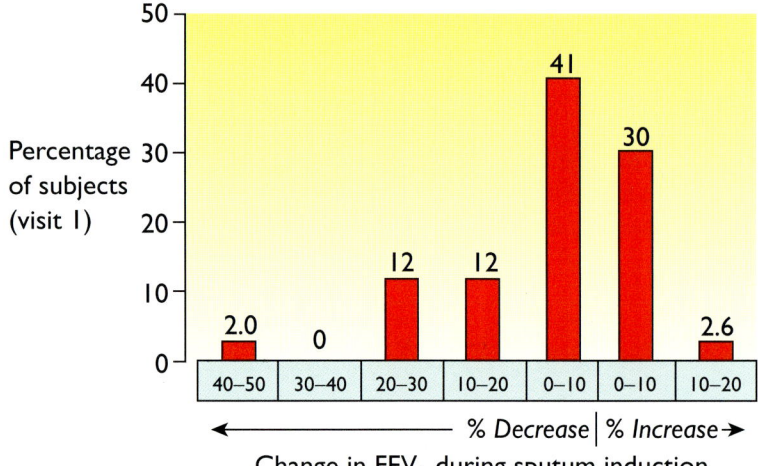

Figure 8.9 Effects of sputum induction on lung function observed in a US-based multicentre study. The figure shows the percentage of 79 subjects studied who had an increase or decrease in their FEV_1 from the post-bronchodilator baseline during sputum induction. Reproduced with permission from Fahy JV, Boushey HA, Lazarus SC, *et al.* Safety and reproducibility of sputum induction in asthmatic subjects in a multicenter trial. *Am J Respir Crit Care Med* 2002;163: 1470–5

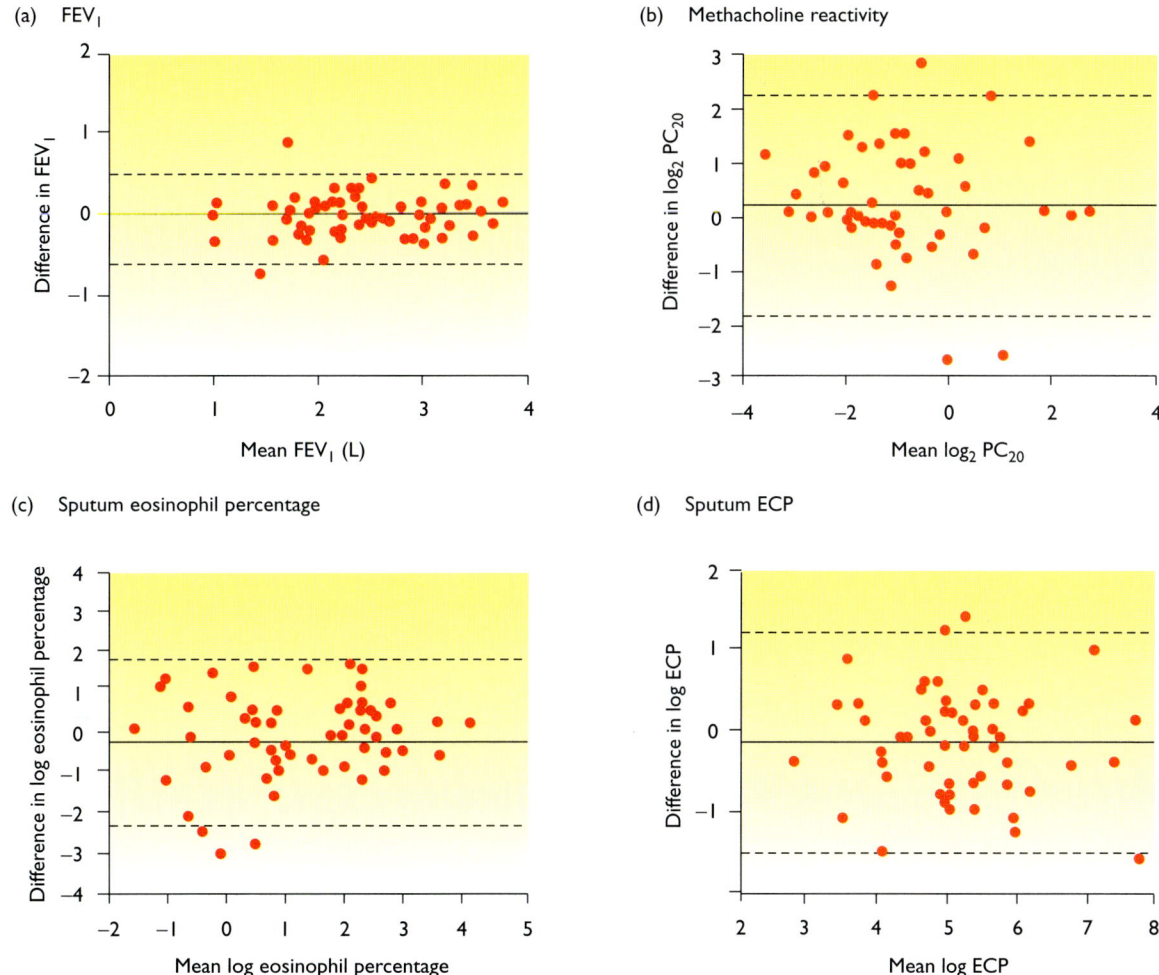

Figure 8.10 Bland–Altman summaries for the reproducibility of eosinophil percentages and eosinophil cationic protein (ECP) levels in induced sputum compared with the reproducibility of FEV₁ and methacholine PC₂₀ in the same subjects in a US-based multi-centre trial. The Bland–Altman summaries plot the difference in the measures between the two visits versus the average value of the measures for the two visits for each subjects. The overall average difference (solid line) ± 2 standard deviations (dashed line) is also shown in the plot. Adapted with permission from Fahy JV, Boushey HA, Lazarus SC, *et al.* Safety and reproducibility of sputum induction in asthmatic subjects in a multicenter trial. *Am J Respir Crit Care Med* 2002;163:1470–5

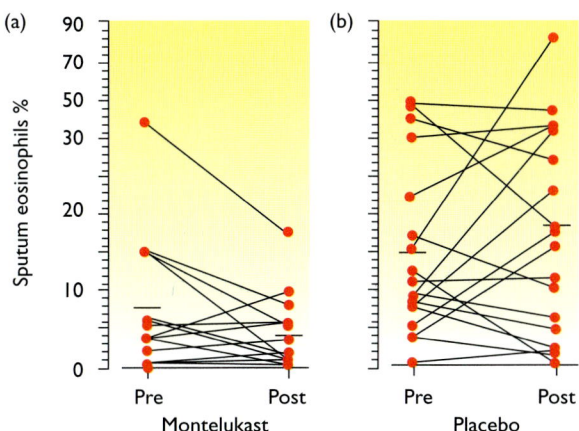

Figure 8.11 The effect of (a) montelukast (n = 16, dose 10 mg/day) and (b) placebo (n = 20) on sputum eosinophils at baseline (Pre) and after 4 weeks (Post) of therapy in a six-centre trial. Each point is the mean from two blinded readers. Horizontal bars represent mean values. The percentage of eosinophils (mean ± SD) decreased form 7.53 ± 9.52 to 3.88 ± 4.67 in the montelukast group and increased from 14.54 ± 14.40 to 17.90 ± 19.79 in the placebo group (*p* = 0.026). Reproduced with permission from Pizzichini E, Leff JA, Hargreave FE, *et al.* Montelukast reduces airway eosinophilic inflammation in asthma: a randomised, controlled trial. *Eur Resp J* 1999;14:12–18

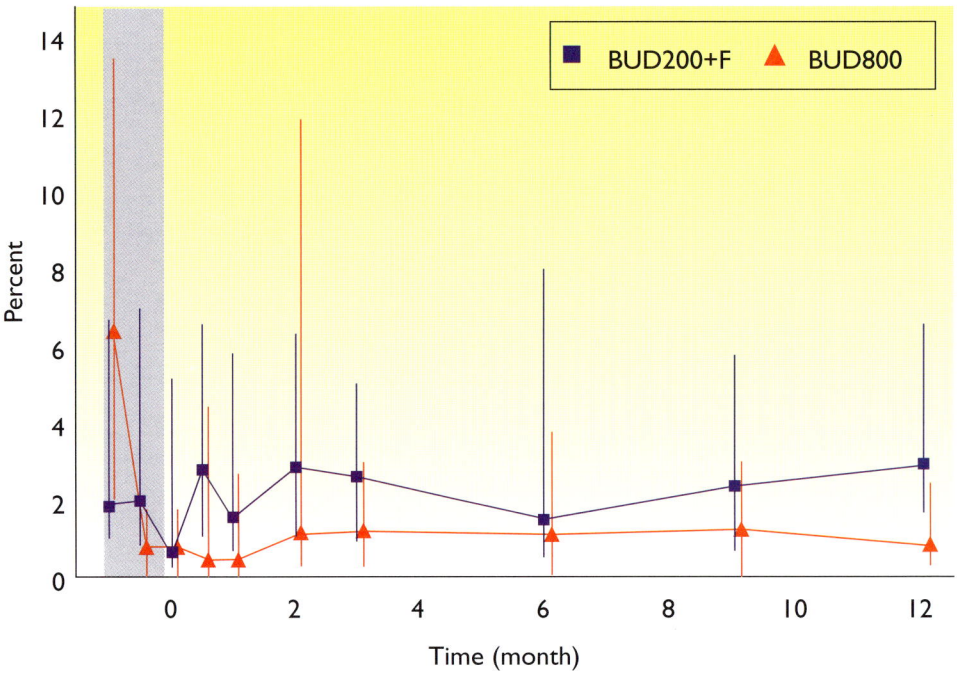

Figure 8.12 Proportion of eosinophils in induced sputum expressed as a percentage of total nucleated nonsquamous cells, during the run-in (shaded area) and treatment period in a three-centre trial. Values are expressed as median ± interquartile range. During the run-in period, all patients were treated with budesonide 800 μg, twice daily. Patients were then randomized to treatment with budesonide 100 μg plus formoterol 12 μg, or budesonide 400 μg, twice daily. A significant decrease in eosinophils was observed over the 1-month run-in period. However, during the randomized treatment period, no significant changes occurred either within or between both treatment groups, suggesting that treatment with a lower dose of corticosteroid in combination with a long-acting β_2-agonist did not cause a deterioration in eosinophilic inflammation when compared with the higher dose of corticosteroid. Reproduced with permission from Kips JC, O'Connor B, O'Byrne PM, *et al*. A long-term study of the antiinflammatory effect of low-dose budesonide plus formoterol versus high-dose budesonide in asthma. *Am J Respir Crit Care Med* 2000;161:996–1001

Table 8.1 Sample size estimates for sputum eosinophil, sputum eosinophil cationic protein (ECP) and sputum tryptase for different effect sizes and power based on an US-based multi-centre study. Reproduced with permission from Fahy JV, Boushey HA, Lazarus SC, *et al*. Safety and reproducibility of sputum induction in asthmatic subjects in a multicenter trial. *Am J Respir Crit Care Med* 2002;163:1470–5

	Effect size								
	Sputum eosinophil percentage			Sputum ECP			Sputum tryptase		
Power	25%	50%	75%	25%	50%	75%	25%	50%	75%
	I. One sample (one treatment)*								
0.8	179	52	28	71	21	11	89	23	10
0.9	240	69	37	95	28	15	119	30	14
	II. Two sample (treatment and placebo)**								
0.8	554	108	32	238	46	14	416	70	16
0.9	742	144	42	318	162	20	556	94	20

The sample size estimates are based on the variability in the outcomes over the two-visit study; *effect size is relative change over time; **effect size is difference in relative change over time standardized protocol for sputum induction and sample processing and adhere to the same safety measurements. This is achieved by providing adequate training for staff who will conduct the procedure and centralizing analysis in one center. If such an approach is taken, risks of sputum induction are minimal and acceptable (Figure 8.9), with the majority of subjects (70%) experiencing a change in FEV_1 that is within 10% of baseline. The reproducibility of measures of inflammation in induced sputum compares favorably with the reproducibility of other commonly used outcome measures of asthma control (Figure 8.10).

Induced sputum has been used to assess the anti-inflammatory effect of montelukast, a cysteinyl leukotriene receptor antagonist, on airway eosinophils in a six-center, placebo-controlled, parallel-group trial of 4 weeks (Figure 8.11). In this study also, all the technicians were trained centrally to ensure sputum induction and processing consistency and a central laboratory was used for reading the sputum slides.

Induced sputum has been validated as a marker of airway inflammation in a number of short-term studies. Similarly, induced sputum was analyzed in a three-center 1-year follow-up study assessing the effect of two treatment strategies on moderate asthma (Figure 8.12). The study illustrated that treatment of moderate asthma for 1 month with a high dose of inhaled budesonide significantly reduced activated eosinophil numbers in induced sputum, and that no significant differences in sputum markers of airway inflammation were observed during the ensuing 1-year treatment period with a low dose of budesonide plus the long-acting β_2-agonist, formoterol, compared with a four-fold higher dose of budesonide. This study provided useful information concerning the use of this technique as an outcome measure in long-term clinical trials.

Several studies have demonstrated that there can be a wide range of numbers of sputum eosinophils in patients with asthma that appears well controlled according to clinical criteria. This has impact on the study design, as this clinically undetectable variability in sputum eosinophil counts could result in significant differences at randomization. The variability can also affect the power of the study and increase the number of patients needed to obtain significant effects (Table 8.1). An important question also relates to the pathophysiological significance of a given eosinophil count in sputum, and what would be a clinically significant reduction in sputum eosinophil numbers. One approach could be to try to lower the percentage of sputum eosinophils to close to normal cell counts (approximately 2.5%). Based on published data, the European Respiratory Task Force on Induced Sputum has suggested that a reduction in eosinophil counts of at least 50% was clinically relevant.

One important question relates to patients' inclusion criteria for a clinical trial. Depending on the main outcome it might be acceptable to select patients based on the number of eosinophils in their sputum samples. However, selecting patients with a set degree of sputum eosinophila might lead to a selection bias that would make it more difficult to extrapolate the results to the general asthmatic population. On the other hand, the degree of sputum eosinophilia can be used to identify subtypes of patients to demonstrate treatment effects as was done in one study of COPD patients (see Figures 9.13 and 9.14 in Chapter 9).

In conclusion, induced sputum has been shown to be very useful in helping demonstrate and elucidate anti-inflammatory effects of drugs. Future studies will need to establish more precisely what are the clinically relevant changes in the various inflammatory mediators in induced sputum following treatment.

CHAPTER 9

The use of induced sputum in clinical practice

Ian D. Pavord, Debbie Parker and Ruth H. Green

Induced sputum is primarily a research tool. However, recent studies that focused on questions of direct clinical relevance have begun to identify a potential role for this technique in the everyday management of respiratory diseases. Based on current evidence, the use of induced sputum can be recommended for uncertain diagnosis of asthma, and chronic cough. Emerging evidence from studies in asthma suggests that repeated analysis of induced sputum might be able to improve asthma management by reducing the frequency of exacerbations.

Asthma is usually easily diagnosed on the basis of typical symptoms and reversible airflow limitation. Occasionally, this is not the case, which can lead to incorrect labeling of a patient's symptoms as asthma and further mismanagement. One study has shown that 160 of 263 subjects referred to a tertiary referral center with suspected asthma received an alternative diagnosis. Many of these had received prolonged treatment with potentially toxic therapy before the correct diagnosis was reached. Patients suspected by non-specialists to suffer from asthma who are subsequently shown by specialists to have pseudoasthma are diagnosed with a variety of conditions including rhinitis, gastroesophageal reflux, hyperventilation syndrome, obstructive apnea, chronic bronchitis, post-viral cough, unexplained dry cough and eosinophilic bronchitis (Table 9.1).

Making the distinction between asthma and pseudoasthma is an important problem in secondary and tertiary care. Sputum eosinophilia, as the most characteristic cellular abnormality in asthma, can be used to help diagnose asthma when the history is unreliable, and the evidence of reversible airflow limitation is either insufficient or absent. One study has shown that in adults with asthma, who have normal or near-normal spirometric values, methacholine challenge (i.e. methacholine PC_{20}) and the sputum differential eosinophil count are the most valid tests and are, by implication, the most clinically useful for discriminating patients with asthma from those with pseudoasthma (Table 9.2 and Figure 9.1).

The acute bronchodilatory response to inhaled salbutamol and the peak expiratory flow (PEF) variability, derived from twice-daily PEF readings, were no more valid than the identification of an obstructive spirogram for this purpose. The patients with pseudoasthma had a range of conditions that was similar to that described above, and, thus, they are likely to be representative of a wider population with conditions that are commonly confused with asthma.

It important to note that the extent of eosinophilic airway inflammation, as reflected by this sputum eosinophil count, is not closely related to the presence of symptoms or abnormal airway function. Thus, some patients (one example is

Table 9.1   Clinical features, investigation, and treatment of various causes of pseudoasthma. Adapted with permission from Hunter CJ, *et al.* A comparison of the validity of different diagnostic tests in adults with asthma. *Chest* 2002;121:1051–7

Rhinitis

Rhinorrhea, nasal obstruction, sinus pain, sneezing, nasal itch, postnasal drip

Nasal secretions, nasal or pharyngeal mucosal inflammation

Sinus radiograph/CT scan showing mucosal thickening and/or fluid level

Topical budesonide/beclomethasone 100 µg twice daily; in selected cases, topical ipratropium bromide (40 µg twice daily), topical xylometazoline, oral antihistamine

Gastroesophageal reflux

Heartburn, flatulence, water brash

Barium swallow, endoscopy, and 24-h esophageal manometry and pH in selected cases

Weight reduction, elevation of head of bed, no eating within 2 h of bedtime, acid suppression; prokinetic agent in selected cases

Chronic bronchitis

Productive morning cough > 3 months/year for > 1 year; smoking history

Coarse crackles

Stop smoking

Postviral cough

Onset following viral upper respiratory tract infection

Normal

Observation

Hyperventilation syndrome

Sudden dyspnea occurring at rest, associated paresthesias, dizziness

Normal

20 deep breath test

Rebreathing exercises, relaxation, identification of triggering situations

Obstructive sleep apnea

Daytime somnolence, unrefreshed sleep, snoring

Sleep study

Continuous positive-airway pressure

Unexplained dry cough

Persistent nonproductive cough

Normal

Antitussives

---

shown in Figure 9.2) have apparently well controlled asthma with normal PEF, normal PEF variability, no symptoms and minimal requirement for rescue β_2-agonist use, despite marked sputum eosinophilia. In contrast, other patients (an example is shown in Figure 9.3) have poorly controlled symptoms and marked PEF variability, with a need for frequent rescue β_2-agonist but no sputum evidence of eosinophilic airway inflammation. Such patients might benefit from increased bronchodilator use.

Interestingly, the first patient has a history of three life-threatening attacks of asthma. These cases illustrate the disassociation between eosinophilic airway inflammation and other features of the asthmatic state and emphasize that one can draw few inferences about lower airway inflammation from a routine clinical assessment.

There is growing evidence that a sub-group of patients with symptomatic asthma have no sputum evidence of eosinophilic airway inflammation

Table 9.2 Sensitivity and specificity of spirometry, bronchial hyperresponsiveness, and sputum and blood eosinophilia in the diagnosis of asthma. Spirometry values are given as mean (SEM). PEF values recorded in the morning as a percentage of mean of morning and evening values (A%M), blood and sputum eosinophil counts, and PC$_{20}$ are given as geometric mean (log SEM), BD, bronchodilator. Adapted with permission from Hunter CJ, Brightling CE, Pavord ID, *et al*. A comparison of the validity of different diagnostic tests in adults with asthma. *Chest* 2002;121:1051–7

Test	Normal range	Sensitivity (%)	Specificity (%)
FEV$_1$: FVC ratio	> 76.6%	61	60
BD response	< 2.9%	49	70
PEF A%M	< 21.6%	43	75
PC$_{20}$	> 8 mg/mL	91	90
Sputum eosinophil count	< 1%	72	80
Blood eosinophil count	< 6.3%	21	100

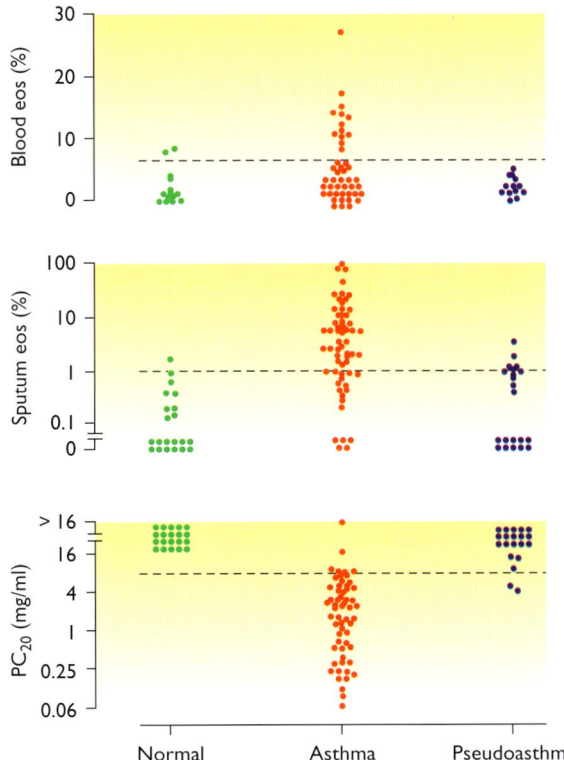

Figure 9.1 Individual measurements of blood and sputum eosinophil (eos) counts and methacholine responsiveness (assessed as PC$_{20}$) in healthy control subjects (normal), subjects with asthma and subjects with pseudoasthma. Hashed lines represent limits of normal for the three measurements. Reproduced with permission from Hunter CJ, Brightling CE, Pavord ID, *et al*. A comparison of the validity of different diagnostic tests in adults with asthma. *Chest* 2002;121:1051–7

(Figure 9.4). This phenotype may be particularly prevalent in more severe asthma but is also seen in patients with milder disease who are not taking inhaled corticosteroids. Many patients have a sputum neutrophilia suggesting a fundamental difference in the nature of the lower airway inflammatory response. Uncontrolled studies have

shown that patients with non-eosinophilic asthma respond less well to treatment with inhaled corticosteroids than those with eosinophilic asthma. However, until controlled studies have shown that non-eosinophilic asthma does not respond favorably to corticosteroids, standard asthma treatment guidelines should be adhered to.

WEEK 2 DATE:	12/10	13/10	14/10	15/10	16/10	17/10	18/10
Daytime asthma:	0	0	0	0	0	0	0
Night-time wakening :	0	0	0	0	0	0	0
PEAK FLOW AM	440	440	440	440	430	430	430
Best of 3 PM	500	490	470	470	470	470	460
Number of puffs of Ventotin or Bricanyl per 24 hours	2	2	2	2	2	2	2

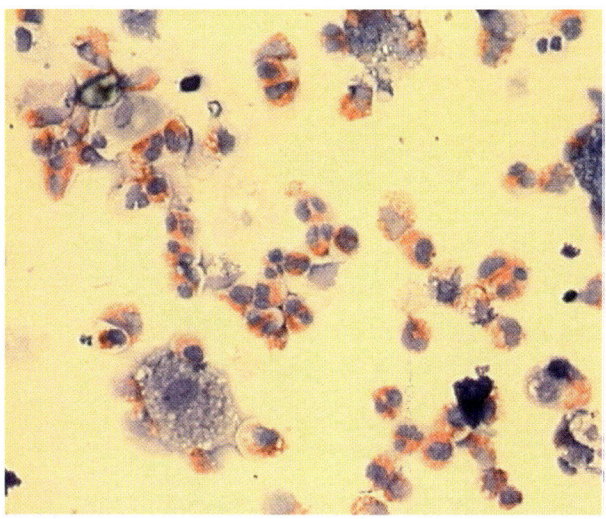

79% eosinophils

Figure 9.2   Diary card recordings (top) and induced sputum cytospin preparation from a 42-year-old asthmatic with severe corticosteroid-dependent but currently stable asthma who had been previously ventilated on three occasions. Despite apparently good clinical control at the time of study, this patient has markedly raised numbers of eosinophils in the sputum (note the eosinophilic staining of the cytoplasm and characteristic bi-lobed nucleus)

The disassociation between symptoms/airway dysfunction and exacerbations/eosinophilic airway inflammation has important implications for the way asthma is managed (Figure 9.7). There is increasing evidence that eosinophilic airway inflammation is more closely related to asthma exacerbations than day-to-day symptoms and airway dysfunction. One implication is the suggestion that measuring airway inflammation and modifying treatment to minimize eosinophilic airway inflammation as well as control symptoms and maximize lung function might result in better control of asthma exacerbations than a traditional management approach. This hypothesis has been tested in a randomized controlled trial of 74 subjects with moderate to severe asthma attending as outpatients. Asthmatics were randomized to be treated either according to the British Thoracic Society (BTS) recommended guidelines or a management strategy which suggested adjusting treatment according to the presence of raised or normal sputum eosinophil counts. In the sputum management group, decisions about anti-inflammatory treatment were made in accordance with an algorithm, based on maintenance of a sputum eosinophil count below 3% with a minimum dose

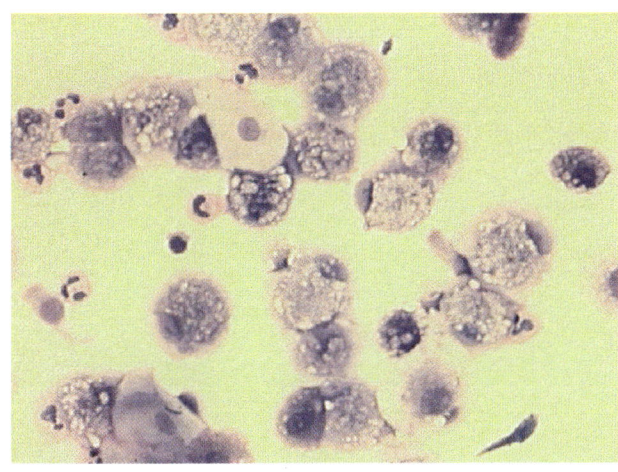

0% eosinophils

WEEK 1 DATE :	22–11	23	24	25	26	27	28
Daytime asthma:	2	2	2	2	3	1	1
Night-time wakening :	3	3	2	3	3	2	3
PEAK FLOW AM	290	210	200	300	280	350	400
Best of 3 PM	300	250	230	250	200	350	350
Number of puffs of Ventolin or Bricanyl per 24 hours	11	15	13	12	16	8	6

Figure 9.3 Diary card recordings (top) and induced sputum cytospin preparation from a 22-year-old female asthmatic with unstable asthma despite use of inhaled corticosteroids. The majority of cells seen are macrophages. Although this patient does not have sputum eosinophilia she has unstable asthma

of anti-inflammatory treatment. The cut-off of 3% was chosen because this was previously shown to identify individuals with corticosteroid-responsive asthma. If the sputum eosinophil count was less than 1%, anti-inflammatory treatment was reduced irrespective of asthma control. If the eosinophil count was 1–3%, no changes to anti-inflammatory treatment were made, and if the eosinophil count was greater than 3%, anti-inflammatory treatment was increased. Decisions about changes in bronchodilator treatment were based on individual patients' symptoms, PEF readings, and use of rescue β_2-agonists compared with baseline – using the same criteria as in the BTS management group. Management decisions were made by an independent individual who was unaware of the clinical characteristics of the patient, and who recorded separate treatment plans to be followed depending on whether the patient's asthma was controlled well or poorly.

The strategy based on sputum eosinophil counts achieved significantly better control of eosinophilic airway inflammation over the 12-month period of the trial (Figure 9.5). There was a similar improvement in PC_{20}. Both management strategies achieved equivalent control of symptoms, quality of life and disordered airway function (Figure 9.6). However, in the sputum management group there was a marked reduction in severe asthma

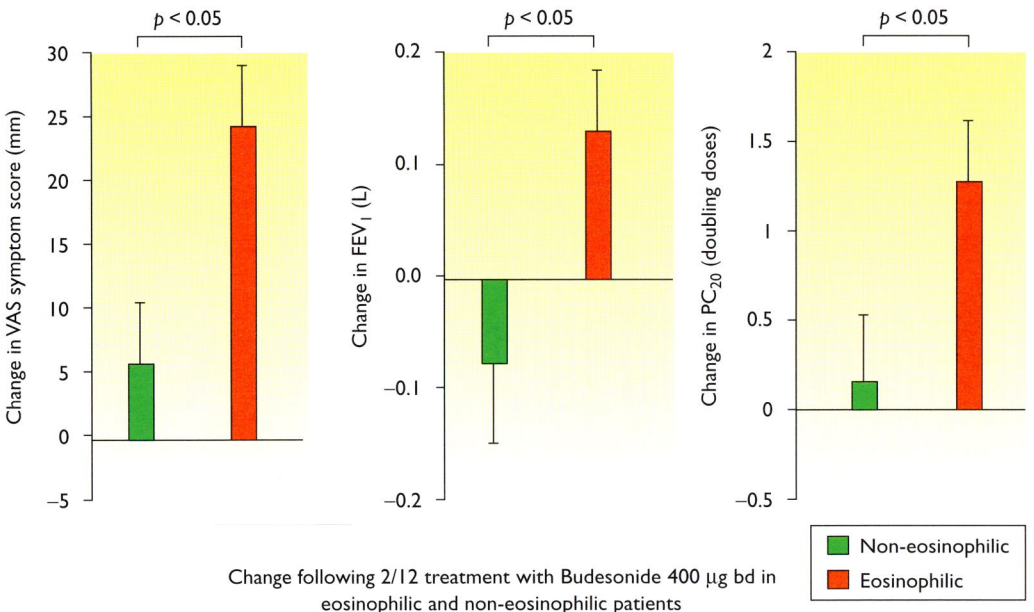

Change following 2/12 treatment with Budesonide 400 μg bd in eosinophilic and non-eosinophilic patients

Figure 9.4   Difference in effect of treatment with the inhaled corticosteroid, budesonide, in asthmatics with and without sputum eosinophilia (sputum eosinophil count > 1.9%). Asthmatics with eosinophilia were found to respond more favourably to budesonide in respect of symptoms, lung function (FEV_1) and bronchial responsiveness (as shown by PC_{20} methacholine). VAS, visual analog score. Reproduced with permission from Green RH, Brightling CE, Pavord ID, *et al*. Analysis of induced sputum in adults with asthma: identification of subgroup with isolated sputum neutrophilia and poor response to inhaled corticosteroids. *Thorax* 2002;57:875–9

exacerbations and significantly fewer hospital admissions with asthma exacerbations (Figure 9.7). This study therefore supports the view that assessment of airway inflammation provides additional important information, which is necessary for the optimum management of more severe asthmatics.

INDUCED SPUTUM AS A TOOL FOR DIAGNOSING COUGH

One of the more interesting observations with induced sputum is that sputum eosinophilia can be seen in patients with chronic cough in the absence of any of the abnormalities of airway function that characterize asthma (Figure 9.8). This condition has been called eosinophilic bronchitis. Eosinophilic bronchitis can only be positively identified in the clinic if airway inflammation is measured in sputum. Unlike other types of cough, such as that associated with reflux or post-viral cough, this type

of cough responds favorably to treatment with inhaled corticosteroids.

Cough is a very common complaint in general practice. Usually it is self-limiting, but occasionally it persists for several months if not years. Various approaches have been taken to manage this condition, ranging from nihilism to complex algorithms. In one study in which induced sputum induction was added to a common investigation algorithm for chronic cough (Figure 9.9), eosinophilic bronchitis emerged as one of the more common causes (Table 9.3). The presence of increased eosinophil counts in induced sputum enables this condition to be diagnosed accurately and treated appropriately. Unlike other causes of cough, such as reflux esophagitis, eosinophilic bronchitis responds well to inhaled corticosteroids (Figure 9.10). However, in contrast to asthma, the improvement in eosinophilic airway inflammation is associated with an improvement in cough and a reduction in

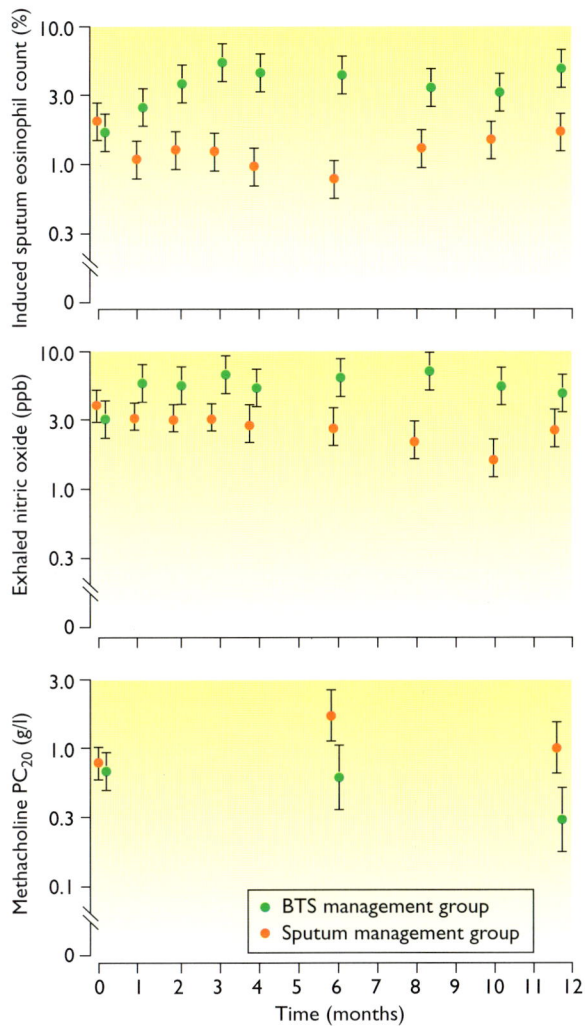

Figure 9.5 Comparison of the effects of two treatment strategies on airway inflammation (induced sputum eosinophil counts and exhaled nitric oxide (NO)) and bronchial hyperresponsiveness (methacholine PC_{20}). One strategy (British Thoracic Society (BTS) management group) utilized standard guidelines of the BTS) and the other (sputum management group) adjusted the anti-inflammatory treatment with corticosteroids based on the eosinophil counts. Reproduced with permission from Green RH, Brightling CE, Pavord ID, *et al*. Asthma exacerbations and sputum eosinophil counts: a randomised controlled trial. *Lancet* 2002;360:1715–21

capsaicin cough sensitivity, but no change in airway responsiveness (Figure 9.11).

USE OF INDUCED SPUTUM IN ASSESSING COPD

COPD is typically associated with an increase in the differential and absolute sputum neutrophil count, although in some patients a sputum eosinophilia is seen. As discussed in Chapter 5, corticosteroids are not very effective in COPD and one reason for this is that these drugs do not inhibit neutrophilic infiltration and activation. A number

of studies have now shown that corticosteroid treatment is more effective in COPD patients with high sputum eosinophil counts; this effect is associated with an improvement in quality of life (CRQ) (Figure 9.12) and forced expiratory volume (FEV_1) (Figure 9.13) but no change in markers of neutrophilic airway inflammation, suggesting that the relatively modest improvements seen with corticosteroid treatment in COPD are due to modulation of one aspect of the lower airway inflammatory response.

In summary, induced sputum has come a long way from being simply a research tool. Increasingly,

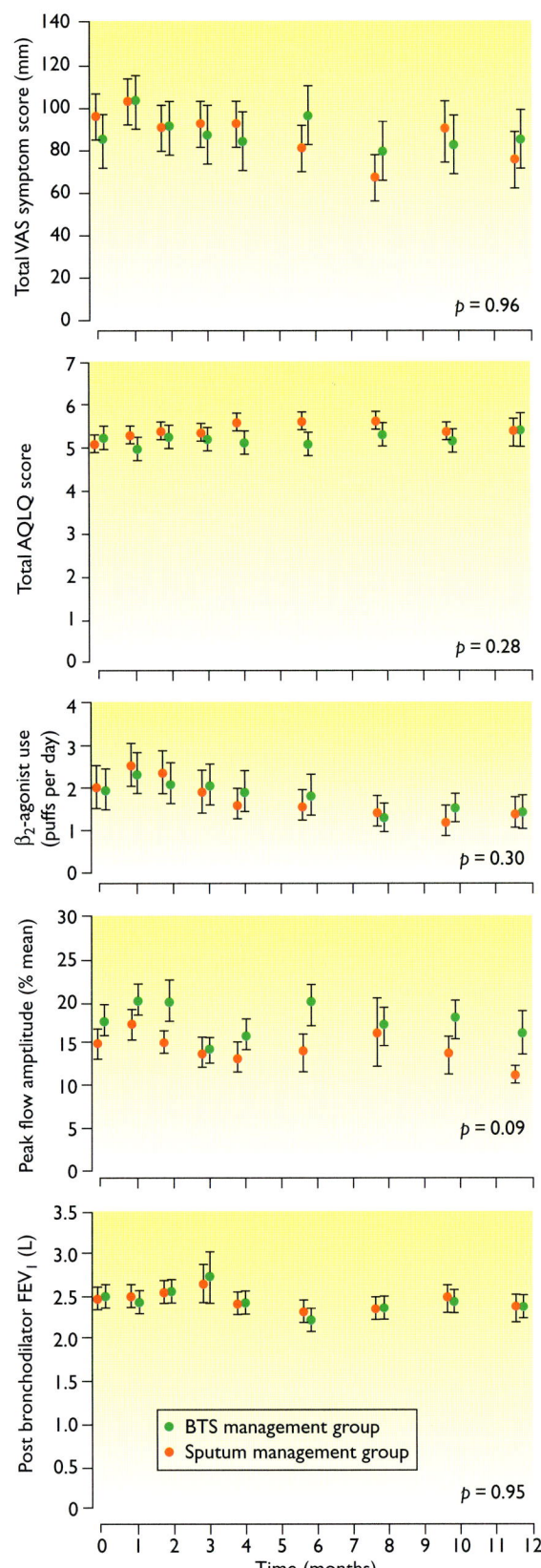

Figure 9.6 Comparison of effects of two treatment strategies on symptom score (assessed by visual analogue score, VAS), asthma quality of life questionnaire (AQLQ) score, β_2-agonist use to relieve asthma symptoms, peak expiratory flow, and post-bronchodilator FEV_1. One strategy (British Thoracic Society (BTS) management group) utilized standard guidelines of the BTS and the other (sputum management group) adjusted the anti-inflammatory treatment with corticosteroids based on eosinophil counts. Reproduced with permission from Green RH, Brightling CE, Pavord ID, *et al*. Asthma exacerbations and sputum eosinophil counts: a randomised controlled trial. *Lancet* 2002;360:1715–21

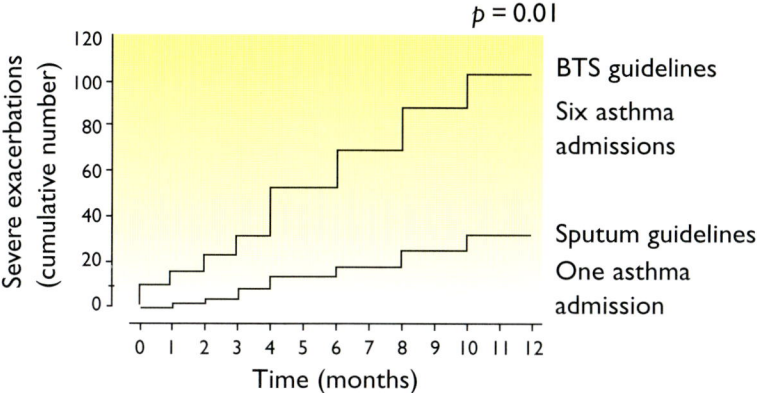

$p = 0.01$

BTS guidelines

Six asthma
admissions

Sputum guidelines

One asthma
admission

Figure 9.7 Comparison of effects of two treatment strategies on rates of severe exacerbations of asthma. One strategy (British Thoracic Society (BTS) guidelines) utilized standard guidelines of the BTS and the other (sputum guidelines) adjusted the anti-inflammatory treatment with corticosteroids based on the eosinophil counts. Reproduced with permission from Green RH, Brightling CE, Pavord ID, *et al*. Asthma exacerbations and sputum eosinophil counts: a randomised controlled trial. *Lancet* 2002;360:1715–21

Asthma

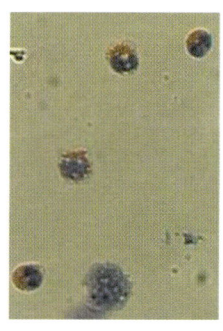

Eosinophilic bronchitis without asthma

- Variable airflow obstruction
- Cough, wheeze, SOB
- Lower-airway hyperresponsiveness
- Sputum eosinophilia

- Chronic cough
- Sputum eosinophilia
- No variable airflow obstruction
- No lower-airway hyperresponsiveness

Figure 9.8 Characteristic features of asthma and eosinophilic bronchitis, some of which are common to both conditions and some which are distinct features of asthma (such as bronchial hyperresponsiveness). Both conditions respond favourably to corticosteroids

applications are being identified for its use in clinical management. Further studies, some of which will involve other inflammatory markers, will almost certainly strengthen the case for this method in routine clinical practice. Some of these are discussed in the next chapter on parenchymal diseases.

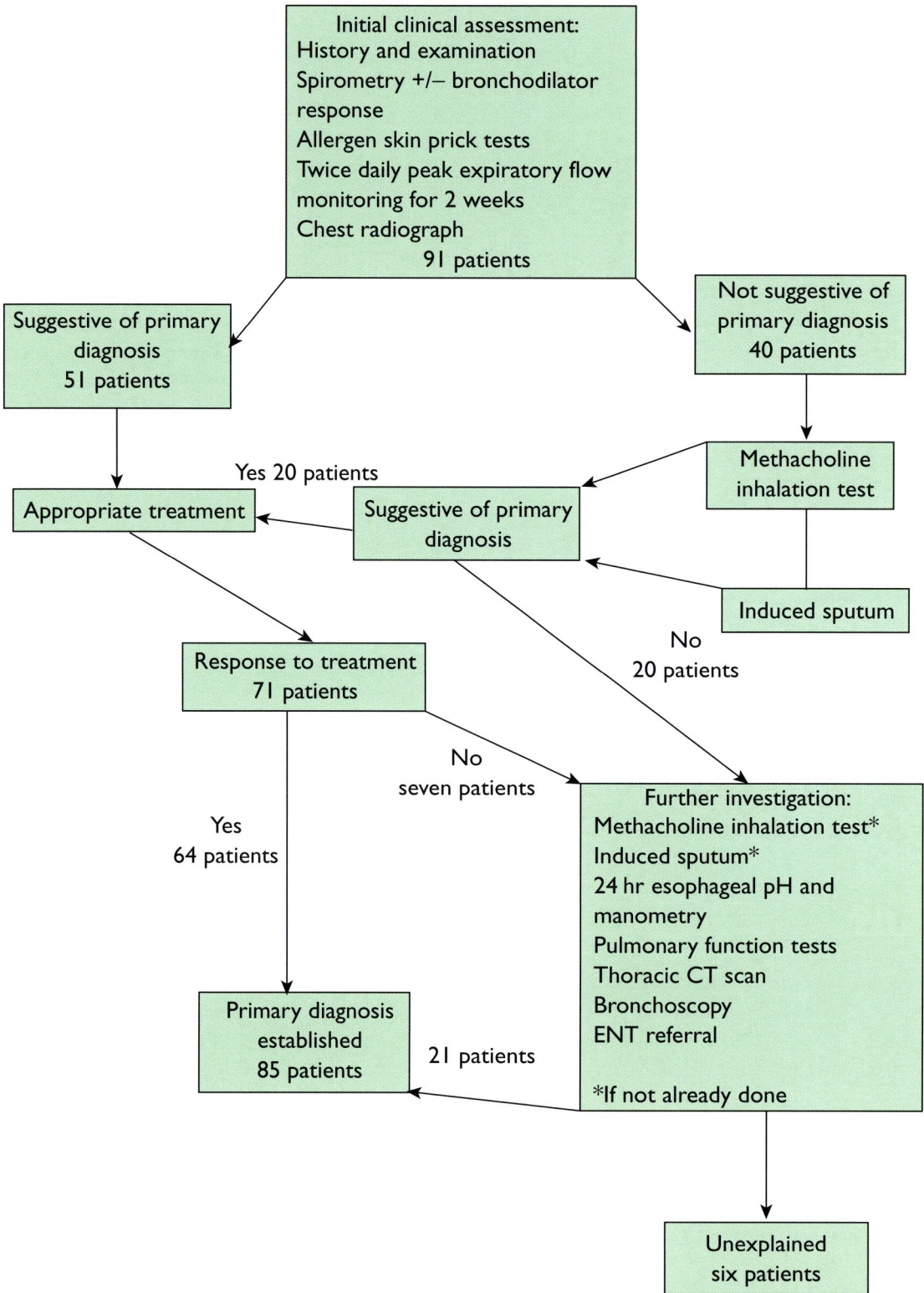

Figure 9.9 Algorithm for investigating chronic cough that includes assessment of induced sputum. The relatively small number of patients for whom no explanation for their cough was found shows that detailed investigation of these difficult patients is rewarding. Reproduced with permission from Brightling CE, Ward R, Pavord ID, *et al*. Eosinophilic bronchitis is an important cause of chronic cough. *Am J Respir Crit Care Med* 1999;160:406–10

Primary cause of chronic cough	Number of patients (%)
Rhinitis/PND	20 (24)
Asthma	16 (17.6)
Post viral	12 (13.2)
Eosinophilic bronchitis	12 (13.2)
Gastroesophageal reflux	7 (7.7)
Unexplained	6 (6.6)
COPD	6 (6.6)
Bronchiectasis	5 (5.5)
ACE inhibitor induced cough	4 (4.4)
Lung cancer	2 (2.2)
Cryptogenic fibrosing alveolitis	1 (1.1)

Table 9.3 Causes of chronic cough. Eosinophilic bronchitis, a novel syndrome characterized by sputum eosinophilia and normal airway responsiveness, is a relatively frequent cause of chronic cough and the application of induced sputum can help diagnose this condition accurately. Reproduced with permission from Brightling CE, Ward R, Pavord ID, *et al*. Eosinophilic bronchitis is an important cause of chronic cough. *Am J Respir Crit Care Med* 1999;160:406–10

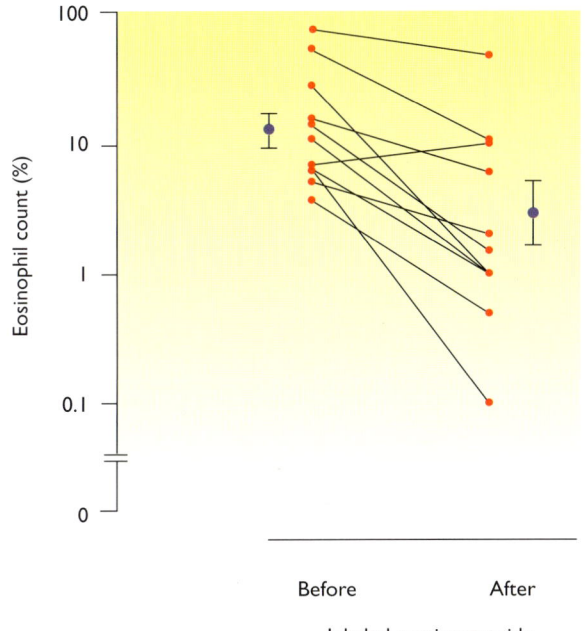

Figure 9.10 The effect of inhaled corticosteroids on eosinophil counts in eosinophilic bronchitis. Reproduced with permission from Brightling CE, Ward R, Wardlaw AJ, Pavord ID. Airway inflammation, airway responsiveness and cough before and after inhaled budesonide in patients with eosinophilic bronchitis. *Eur Respir J* 2000;15:682–6

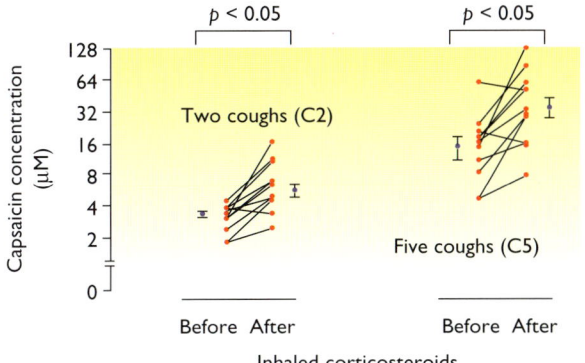

Figure 9.11 The effect of treatment with inhaled corticosteroids on cough induced by capsaicin in patients with eosinophilic bronchitis. The concentrations of capsaicin inducing two and five coughs, respectively are shown before and after treatment. *p* values refer to difference between geometric mean. Reproduced with permission from Brightling CE, Ward R, Wardlaw AJ, Pavord ID. Airway inflammation, airway responsiveness and cough before and after inhaled budesonide in patients with eosinophilic bronchitis. *Eur Respir J* 2000;15:682–6

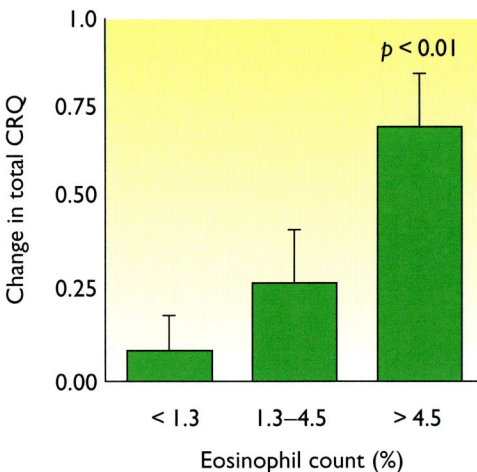

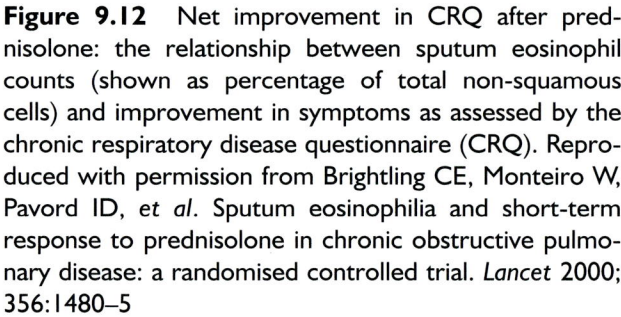

Figure 9.12 Net improvement in CRQ after prednisolone: the relationship between sputum eosinophil counts (shown as percentage of total non-squamous cells) and improvement in symptoms as assessed by the chronic respiratory disease questionnaire (CRQ). Reproduced with permission from Brightling CE, Monteiro W, Pavord ID, *et al*. Sputum eosinophilia and short-term response to prednisolone in chronic obstructive pulmonary disease: a randomised controlled trial. *Lancet* 2000; 356:1480–5

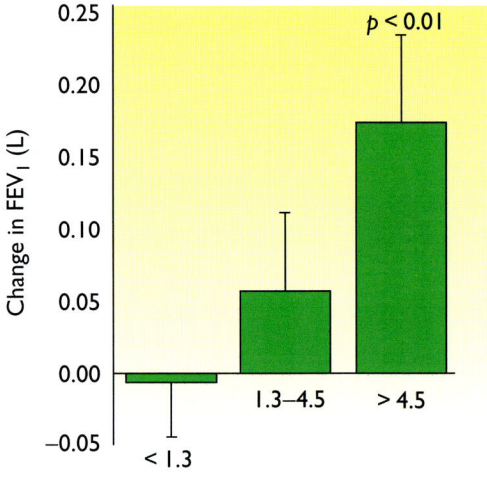

Figure 9.13 Net improvement in FEV₁ after prednisolone: the relationship between sputum eosinophil counts (shown as percentage of total non-squamous cells) and improvement in lung function (FEV₁). Reproduced with permission from Brightling CE, Monteiro W, Pavord ID, *et al*. Sputum eosinophilia and short-term response to prednisolone in chronic obstructive pulmonary disease: a randomised controlled trial. *Lancet* 2000;356:1480–5

CHAPTER 10

Induced sputum in interstitial and occupational lung diseases

Elizabeth Fireman and Ulrich Costabel

The nature and extent of parenchymal lung diseases has, until recently, been assessed by direct, invasive means such as transbronchial biopsy, bronchoalveolar lavage (BAL) or surgical biopsy, or by indirect assessment of symptoms, pulmonary function tests, thoracic imaging or peripheral blood inflammatory markers. Surgical biopsy with tissue diagnosis has been and remains the gold standard, but it has the drawbacks of patient discomfort, the possible risks associated with the procedure itself, and the costs involved. There is an ongoing search for non-invasive techniques that can either substitute the invasive procedures or provide complementary information. One of these alternative techniques is induced sputum analysis. Considerable work has been undertaken to validate the use of induced sputum by comparing it with BAL findings and standardizing the methods to obtain normal values.

OCCUPATIONAL LUNG DISEASES

The assessment of BAL cells and fluid in respect of its particulate burden has been proposed as being a potentially important diagnostic tool in the evaluation of past and present occupational exposure. Although BAL carries only a minimal risk to the patient when applied under proper selection guidelines, it is an invasive procedure and thus not suitable for screening programs, exposure evaluation, or repeated follow-up of workers who have been exposed to hazardous dust. Sputum induction has been compared with BAL for the assessment of exposure to hazardous dust and in the evaluation of patients with pneumoconiosis (silica and hard metal workers) (Figure 10.1). Other studies have been performed in which only induced sputum has been evaluated in subjects suspected of occupational exposure: eosinophil counts have been measured in the induced sputum of asthmatic isocyanate-sensitized subjects and the frequency of bronchial dysplasia has been investigated in the sputum of miners with past exposure. Other researchers have studied the relevance of asbestos bodies, showing that asbestos bodies can be found in spontaneous and induced sputum of exposed workers even years after retirement.

When assessing silica- and hard metal-exposed workers, BAL and induced sputum specimens yield similar quantitative and qualitative results in terms of the number of particles present in samples and the chemical analysis of the particles. Macrophages serve as the major phagocytic cells, which rapidly remove particles to avoid tissue damage. Silica particles can be seen within macrophages in sputum using polarized light and phase contrast microscopy (Figure 10.2).

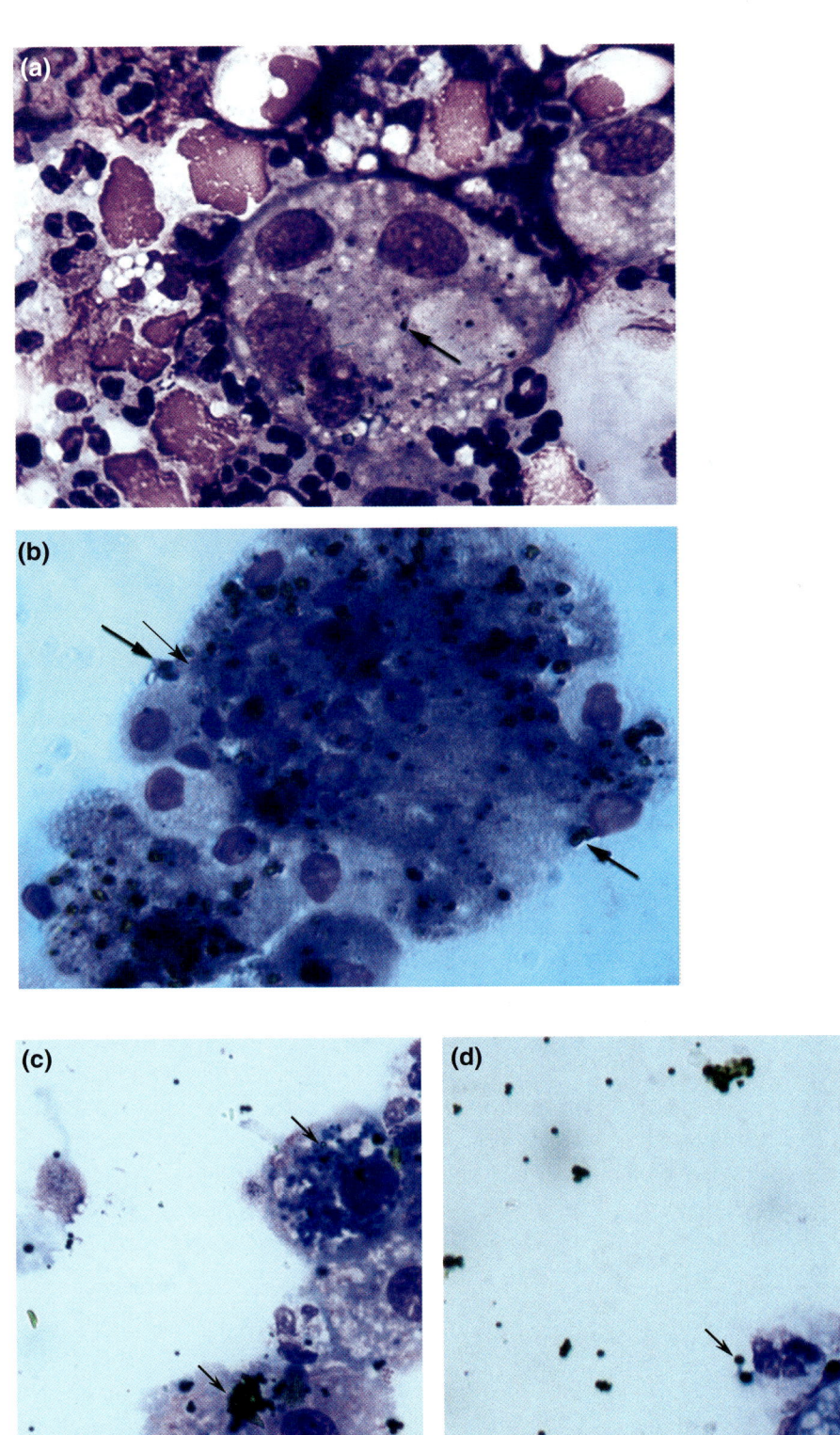

Figure 10.1 Induced sputum in occupational lung disease. Representative sputum samples from heavily exposed workers, with numerous dust particles in the cytoplasm of macrophages – arrows in (a), (b) and (c). Particles can also be seen outside cells, as shown in (d). Giemsa staining, magnification ×100

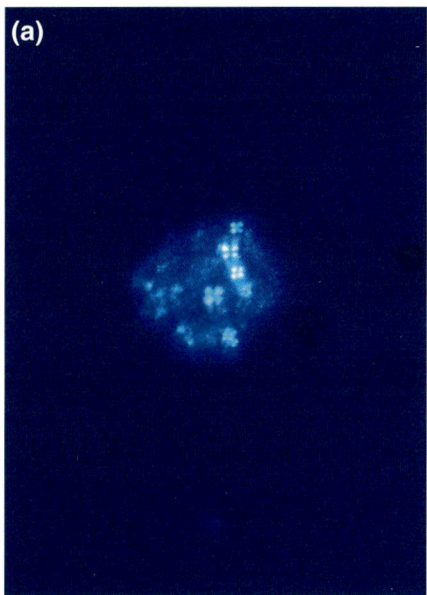
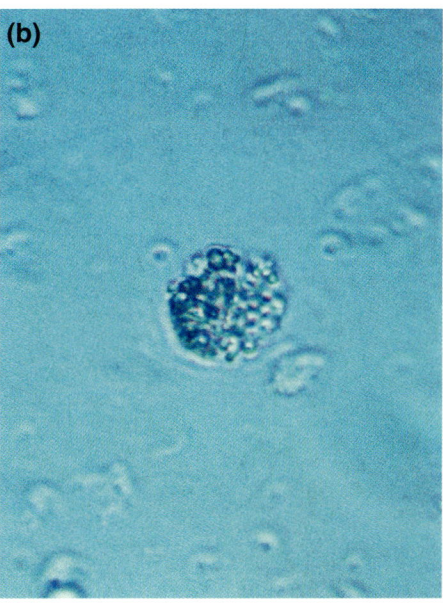

Figure 10.2 Silica particles in macrophages as seen by polarized light (a) and phase contrast (b) microscopy

SARCOIDOSIS AND OTHER GRANULOMATOUS DISORDERS

Several studies have demonstrated higher relative lymphocyte counts (expressed as percentages of total inflammatory cells) in both BAL and induced sputum from patients with sarcoidosis. Percentages of mast cells and eosinophils have been found to be higher in induced sputum samples than in BAL samples in sarcoidosis (Figure 10.3). These data provide evidence to suggest that these cells infiltrate the bronchial mucosa, and support the studies of others showing that patients with sarcoidosis have an increased frequency of bronchial hyperreactivity and signs of increased mast cell activation compared with normal volunteers.

Further analyses and comparison of subpopulations of T lymphocytes in samples obtained by BAL and induced sputum have shown that although the percentages of total CD3+, CD4+ and CD8+ lymphocytes are lower in induced sputum than in BAL, the proportions of these subsets are highly correlated in BAL and induced sputum. These data support the view that the T cell subpopulations present in samples recovered from induced sputum correlate well with those recovered by BAL, and that induced sputum can

effectively identify CD4+ T cell-mediated inflammation and therefore distinguish between sarcoidosis and other non-granulomatous interstitial lung diseases. Moreover, it has been shown that the CD4+/CD8+ ratio and levels of tumor necrosis factor-alpha (TNF-α) are decreased in induced sputum samples after 6 months of treatment.

The differential diagnosis of uveitis is often challenging even for the ophthalmologist. One-third of cases of uveitis are considered related to systemic diseases, with sarcoidosis being one of the most common. Retrospective studies have suggested that isolated uveitis could be an early feature of sarcoidosis. There are, however, no pathognomonic findings for uveitis caused by sarcoidosis. Although BAL is an essentially non-invasive procedure and is a minimal risk for the patient, patients without pulmonary involvement are reluctant to undergo this unpleasant procedure, and there may be ethical problems associated with performing BAL in these patients. Studies have, therefore, been conducted to evaluate sputum induction as a technique for diagnosing sarcoidosis as the cause of uveitis. In one study, 17 patients with uveitis were compared with ten patients with proven sarcoidosis (but with no uveitis) and five healthy volunteers. The patients

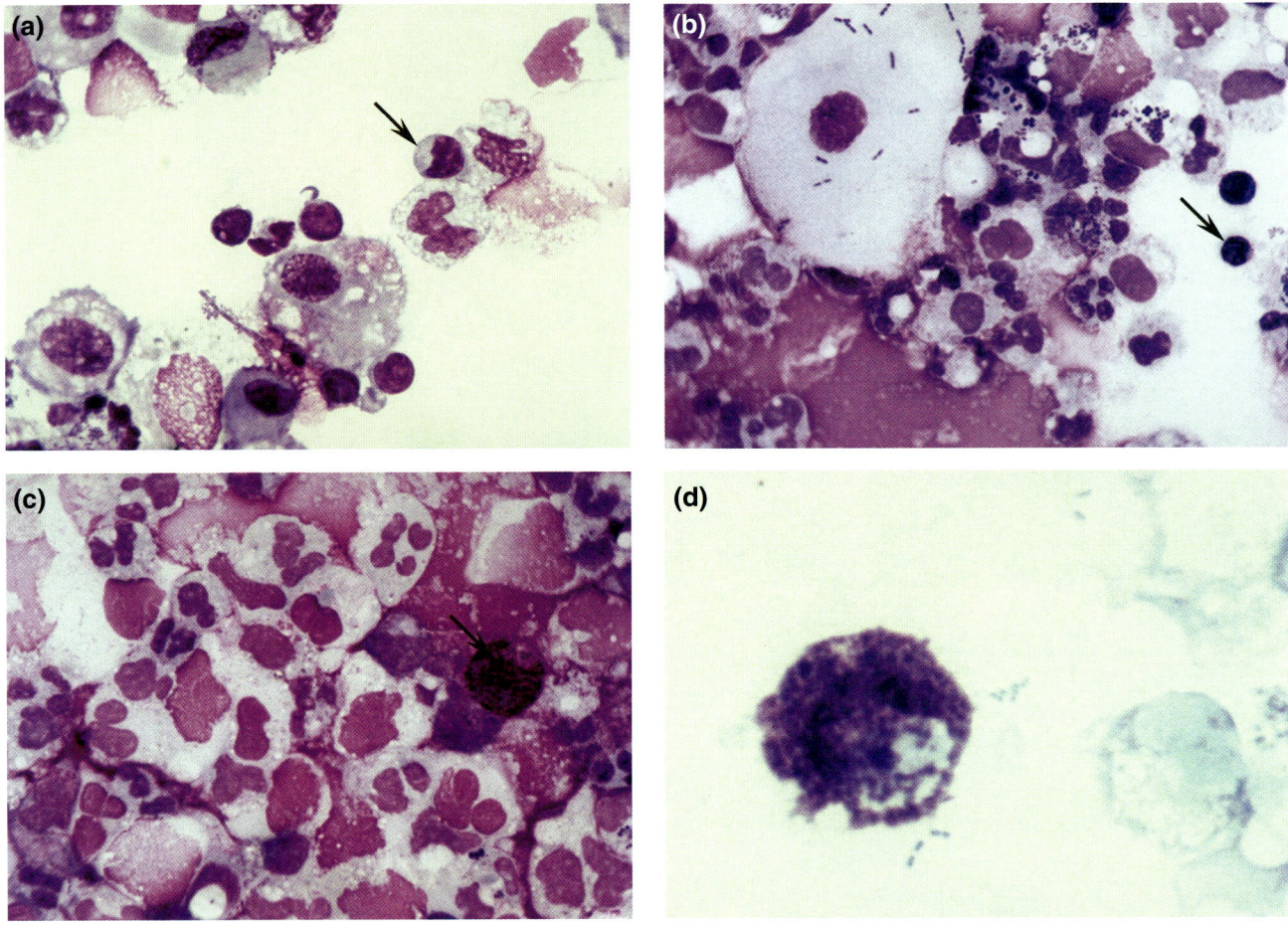

Figure 10.3 Sputum cytology in sarcoidosis. Increased numbers of lymphocytes are a typical finding in sputum of patients with sarcoidosis. (a), (b) and (c), Giemsa staining; (d) toloidine blue staining

with uveitis had no pulmonary symptoms, and their pulmonary function test results were normal. Forty-one percent of these asymptomatic patients had an accumulation of CD4+ cells in their sputum and had elevated angiotensin converting enzyme (ACE) levels similar to patients with proven sarcoidosis and higher than control subjects. A case of a patient with severe anterior uveitis as the only clinical presentation has been reported. This patient had undergone bronchoscopy to rule out lung cancer because of a positive smoking history. Cobblestone changes were seen in the bronchial mucosa and transbronchial biopsies showed abundant non-caseating granulomas. Lymphocyte counts and the CD4+/CD8+ ratio in induced sputum, and ACE levels in serum, were higher than normal (Figure 10.4).

Similarly, pulmonary involvement can be seen in patients with Crohn's disease; this is characterized by accumulation of CD4+ T cells in induced sputum. These findings support the concept of uninterrupted recirculation of CD4+ cells from the mucosa-associated lymphoid tissue, with its activated lymphoid follicles, to effectors sites in the integrated human mucosal immune system.

OTHER INTERSTITIAL LUNG DISEASES

Eosinophils and lymphocytes have been detected in induced sputum samples of patients with other non-granulomatous lung diseases, such as idiopathic pulmonary fibrosis. There is a good correlation between the CD4+/CD8+ subsets recovered from

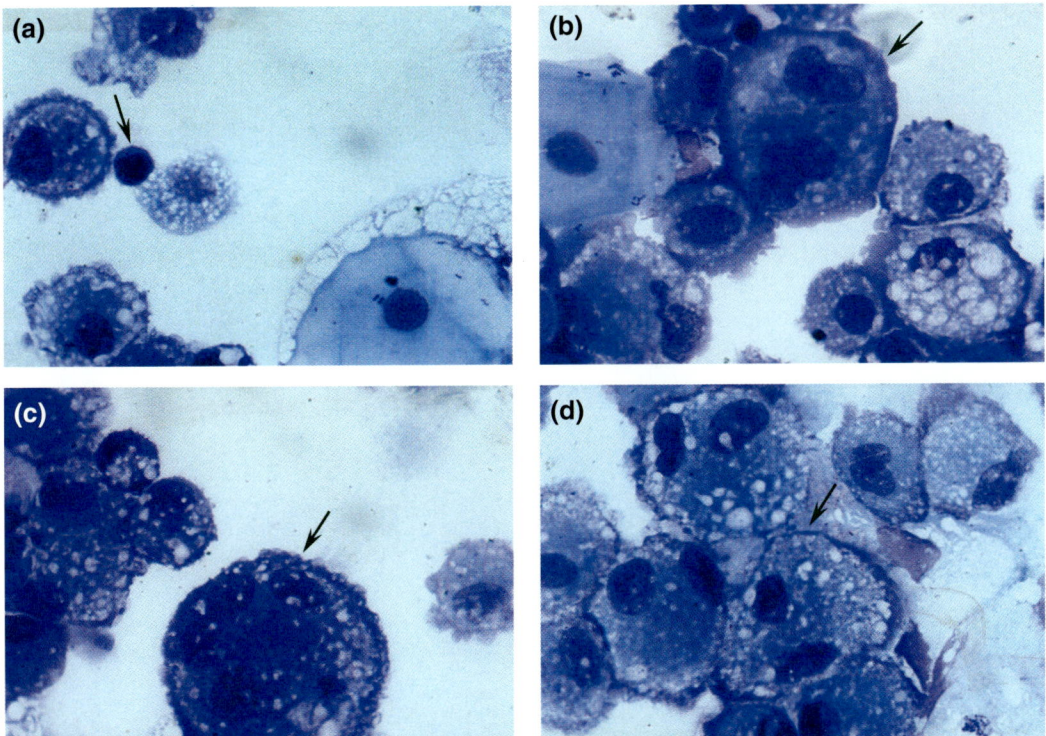

Figure 10.4 Representative induced sputum samples of a patient with isolated uveitis, showing raised lymphocytes (a), macrophages (b) and multinucleated cells (c and d), which are also commonly seen in BAL recovered from sarcoidosis patients

induced sputum and those in BAL in these patients, a finding that may be of crucial importance since a CD4+/CD8+ ratio greater than 1 has been found to have a prognostic value for a favorable response to treatment with corticosteroids.

Macrophages recovered from BAL can be stained with Oil Red O to identify lipid-laden macrophages (LLM). These cells have been shown to be markers of aspiration in parenchymal lung diseases. Since cough is a common and troublesome symptom in these diseases, LLM scores in sputum can be useful as indicators of possible gastro-esophageal reflux syndrome (Figure 10.5).

The presence of phospholipidosis, with foamy cytoplasmic changes in BAL macrophages, can support the diagnosis of pulmonary toxicity caused by the anti-arrhythmic drug, amiodarone. These foamy, vacuolated macrophages can also be detected among alveolar macrophages in the induced sputum of patients receiving amiodarone (Figure 10.6). The presence of these changes, together with the clinical presentation, can help to determine the diagnosis, which is otherwise generally made by exclusion of other causes of parenchymal disease.

Hemosiderin-laden macrophages in induced sputum have been reported to be useful in detecting left ventricular dysfunction (also see Chapter 3). Macrophages containing hemosiderin have also been identified in the induced sputum of patients suffering from diffuse alveolar hemorrhage (Figure 10.7).

Eosinophilic pneumonia is another condition for which analysis of induced sputum can be helpful in respect of management. This inflammatory condition is characterized by a large number of eosinophils in the lungs, usually in the absence of an infectious disease (Figure 10.8).

In conclusion, induced sputum can be considered to be a useful tool for the diagnosis and investigation of parenchymal lung diseases. Its use is likely to increase both for clinical and research purposes; this should allow better evaluation of the usefulness of the technique.

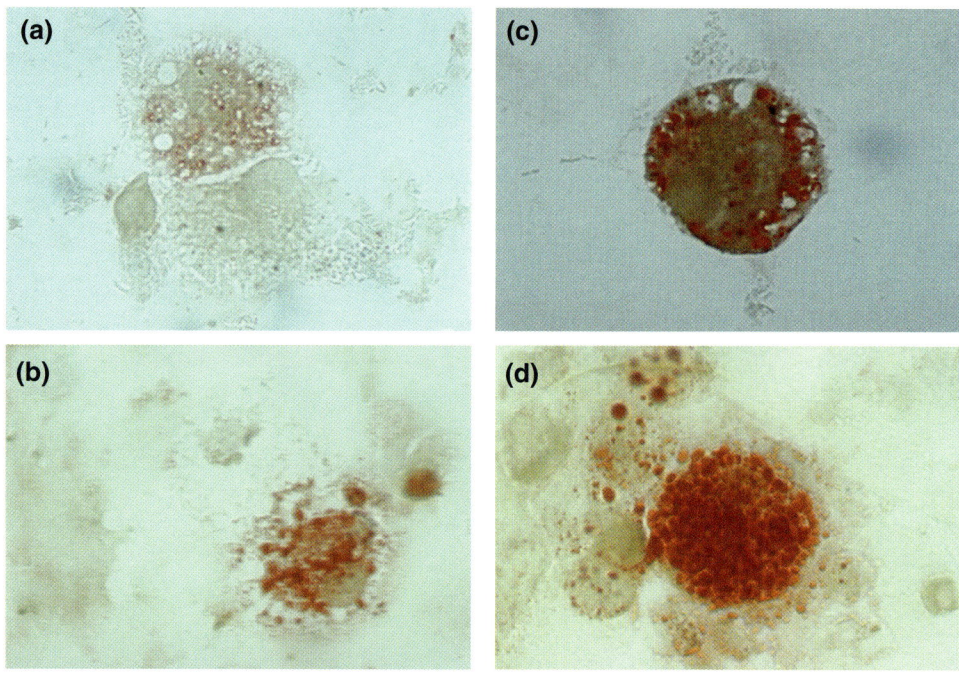

Figure 10.5 Macrophages in sputum stained by Oil Red O. The intensity of staining can vary markedly as seen when comparing a–d

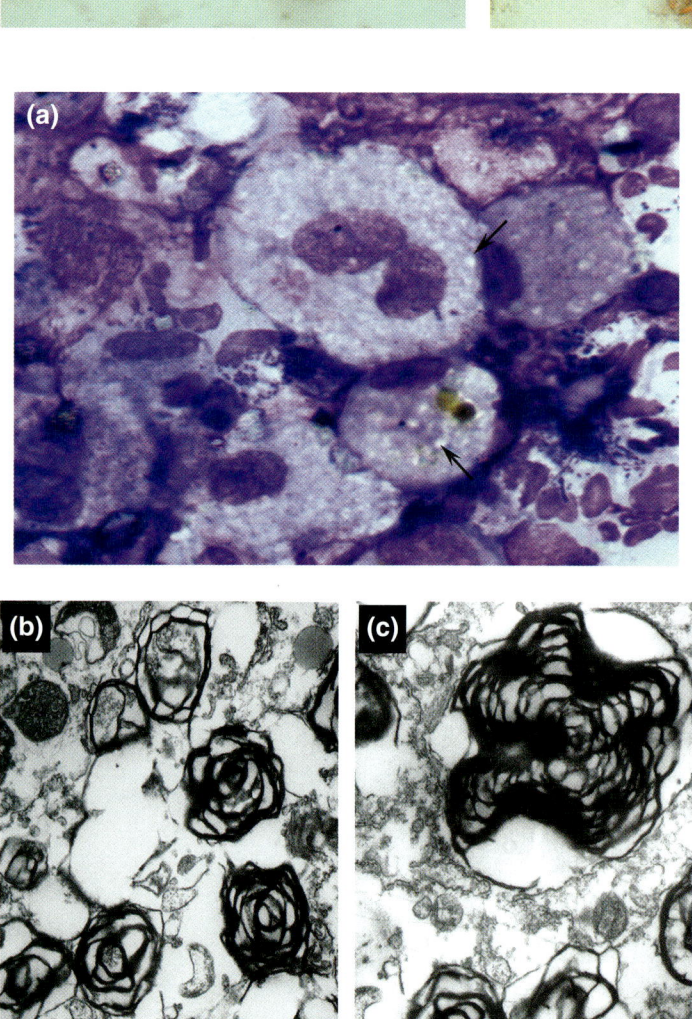

Figure 10.6 Sputum cytology in patients treated with the anti-arrhythmic drug amiodarone. Vacuoles containing phospholipids can be seen by light microscopy in cytospins made from sputum samples stained by Giemsa (a). The phospholipids form lamellar bodies can be identified by electron microscopy (b and c)

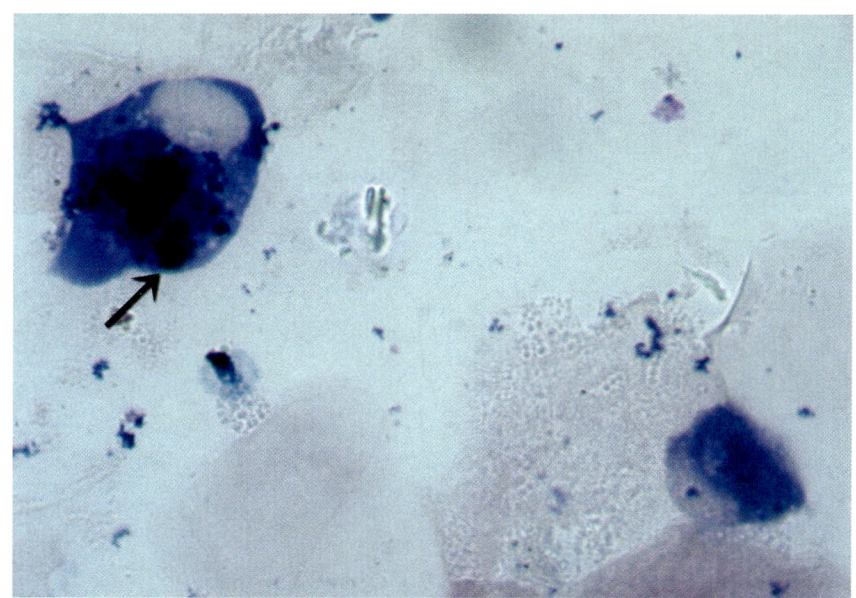

Figure 10.7 Macrophages (arrows) containing hemosiderin can be seen in sputum from patients with diffuse alveolar hemorrhage. This is especially useful in cases where there is microscopic hemorrhage that cannot be easily detected by macroscopic examination of sputum

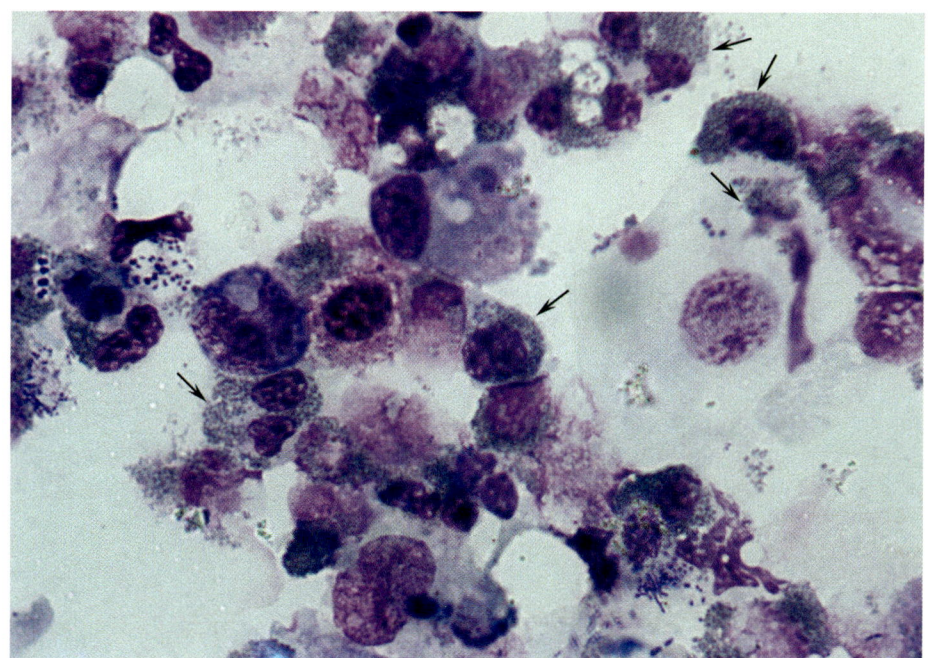

Figure 10.8 Abundant eosinophils (as much as 20–70% of total inflammatory cells) can be seen in induced sputum samples of patients suffering from eosinophilic lung diseases

Subject index

Note: Please note that as the subject of this book is induced sputum, all entries in this index refer to this unless otherwise stated. Page numbers in **bold** represent figures and tables.